Public Administration and Information Technology

Volume 28

Series editor

Manuel Pedro Rodriguez Bolivar, Granada, Spain

More information about this series at http://www.springer.com/series/10796

Yannis Charalabidis • Anneke Zuiderwijk
Charalampos Alexopoulos • Marijn Janssen
Thomas Lampoltshammer • Enrico Ferro

The World of Open Data

Concepts, Methods, Tools and Experiences

Springer

Yannis Charalabidis
Department of Information and Communication Engineering
University of the Aegean
Samos, Greece

Charalampos Alexopoulos
Department of Information and Communication Engineering
University of the Aegean
Samos, Greece

Thomas Lampoltshammer
Department for E-Governance and Administration
Danube University Krems
Krems, Austria

Anneke Zuiderwijk
Faculty of Technology, Policy & Management
Delft University of Technology
Delft, The Netherlands

Marijn Janssen
Faculty of Technology, Policy & Management
Delft University of Technology
Delft, The Netherlands

Enrico Ferro
Head of Innovation Development Department
Istituto Superiore Mario Boella
Turin, Italy

ISSN 2512-1812 ISSN 2512-1839 (electronic)
Public Administration and Information Technology
ISBN 978-3-030-08115-7 ISBN 978-3-319-90850-2 (eBook)
https://doi.org/10.1007/978-3-319-90850-2

Printed on acid-free paper

This Springer imprint is published by the registered company Springer International Publishing AG part of Springer Nature.
The registered company address is: Gewerbestrasse 11, 6330 Cham, Switzerland

Foreword: The Policy View

Public sector information, also referred to as open (government) data, is a valuable resource for the digital economy. It is not only used as valuable raw material for the production of *data-based* services and applications but also brings greater efficiency to the delivery of public services and better informed *decision-making*. Re-use of public sector information promotes citizen empowerment by facilitating government accountability and democratic oversight.[1]

Open data is a fundamental enabler for the *data economy*, linked to the *Digital Single Market* priority of the European Commission[2] about '*bringing down barriers to unlock online opportunities*'. A digital single market should allow the exchange of information, open data included, easily and swiftly across borders, helping citizens and businesses to benefit from them. The ISA^2 programme plays a major role in enabling the cross-border, cross-sector exchange of information.

The ISA^2 programme[3] supports the development of digital solutions that enable public administrations, businesses and citizens in Europe to benefit from interoperable cross-border and cross-sector public services, included in the space of open data.

One of the most important products of the programme is the *European Interoperability Framework* (*EIF*)[4]: This framework offers **recommendations** on

---

[1] Directive 2003/98/EC of the European Parliament and of the Council of 17 November 2003 on the re-use of public sector information – *OJ L 345, 31.12.2003, p. 90–96* (http://data.europa.eu/eli/dir/2003/98/oj)

[2] More information on https://ec.europa.eu/commission/priorities/digital-single-market_en

[3] Decision (EU) 2015/2240 of the European Parliament and of the Council of 25 November 2015 establishing a programme on interoperability solutions and common frameworks for European public administrations, businesses and citizens (ISA^2 programme) as a means for modernising the public sector (text with EEA relevance) – *OJ L 318, 4.12.2015, p. 1–16* (http://data.europa.eu/eli/dec/2015/2240/oj). More information can be found on https://ec.europa.eu/isa2/home_en

[4] Communication from the Commission to the European Parliament, the Council, the European Economic Social Committee and the Committee of the Regions: European Interoperability Framework – Implementation Strategy (COM(2017) 134 final) http://eur-lex.europa.eu/legal-content/EN/TXT/?uri=CELEX:52017DC0134; and also https://ec.europa.eu/isa2/eif_en

how to improve governance of interoperability activities, establish cross-organizational relationships, streamline processes supporting end-to-end digital exchanges, and ensure that both existing and new legislation do not compromise and support interoperability efforts. One of the EIF principles is about openness, *Underlying principle 2*, and there is a whole section on open data and recommendations to public administrations on opening their data.

Last, but not least, as part of the commitment from the ISA^2 programme to open data, it has developed the *DCAT*[5] *Application Profile for data portals* (DCAT-AP)[6]. Due to the increasing number of data portals and the magnitude of available datasets, data users find it difficult to find and access the right data, even if this is available as open data. To ease this issue, the ISA^2 programme has created a common specification, a *common language*, for describing public sector datasets in Europe. This *common language* enables the exchange of descriptions of datasets among data portals. DCAT-AP makes possible for an open data portal, such as the European Data Portal,[7] to easily aggregate descriptions of datasets into a single point of access. The European Data Portal uses DCAT-AP as the common vocabulary for harmonizing descriptions of over 800,000 datasets obtained from 79 data portals of 35 countries. The current version of DCAT-AP has been implemented by 12 countries in Europe. Many more data portals at the European, regional and local level are compliant with DCAT-AP.

We welcome this book and its *future oriented* view on the matter of open data. It will provide invaluable help to public administrations when considering *if* and *how* to publish open data and will also guide both research and policy discussion.

Fidel Santiago
Programme manager for the ISA^2 Programme,
Interoperability Unit, Directorate General for Informatics,
European Commission
Brussels, Belgium

---

[5] DCAT-AP is based on the W3C Data Catalogue Vocabulary (DCAT).

[6] More information on https://joinup.ec.europa.eu/solution/dcat-application-profile-data-portals-europe/about

[7] At http://europeandataportal.eu/

Foreword: The Science View

This book is dedicated to the various aspects and challenges of open data and covers the subject matter comprehensively and demonstrates the diversity of perspectives and approaches, when tackling the issues faced in theory and practice. This book aims at presenting the latest research findings such as theoretical foundations, principles, methodologies, architectures and technical frameworks based on solid and successful cases and lessons learnt from the domain of open data.

Open data is a tremendous resource. It provides the intelligence for insight, invention and exploration that translate into better products and services that improve everyday life and encourage business growth. Research shows that open data has a significant impact in four key areas:

- Improving government
- Empowering citizens
- Creating opportunity
- Solving problems

Open data principles lead to more responsive and smarter government and better service delivery. In order to meet the obligations of the open data movement, agencies must manage data as a strategic asset to be:

- Open by default, protected where required
- Prioritized, discoverable and usable
- Primary and timely
- Well managed, trusted and authoritative
- Free where appropriate
- Subject to public input

The chapters in this book address all important above dimensions and systematically advance our understanding around the open data lifecycle. From policies and organizational issues to data infrastructures and business models, the journey through this book allows the reader to have a systematic, holistic view of the issues and challenges.

I congratulate the authors on the excellent work done and its results. I am certain that this book will be a great commercial and academic success.

Timos Sellis, Fellow IEEE, ACM
Professor, Swinburne University of Technology
Director, Data Science Research Institute
Melbourne, Australia

Foreword: The Industry View

Data is a by-product of the Digital Revolution. It holds an enormous potential in various fields, such as health, food security, climate change, resource efficiency, smart cities and the list goes on. Data has become an asset for growth, innovation and societal resilience. Studies have assessed the size of the potential prize. Whereas the exact numbers may differ from one study to the next, the numbers put forward range from dozens of billions to hundreds of billions. And it is not just about the money. Data-driven products and services improve our everyday life thanks to intelligent transport services or smart energy management for instance.

Today, there is more data and more capacity to manage data and these amounts of data are expected to grow overtime. Data science can be applied to analyse and interpret large amounts of data in order to retrieve valuable insight. This is no longer (only) about structured internal data but about combining data with unstructured internal and external data. Data analytics also witnesses a shift from sample focus groups to exhaustive analysis or 'real' demand without the bias of statistics and forecast inaccuracy.

Numerous European initiatives have been taken in order to reap the benefits of open data. The European Data Portal, launched in 2015 by the European Commission and Member States, is a single point of access to public data resources across Europe. In addition, it offers support to European countries in opening up their data as well as documenting the economic and societal impact of open data. The upcoming Copernicus Data and Information Access Services (DIAS) will take data portals to the next levels by offering access to increasing amounts of Earth Observation data, tools and services. These initiatives materialize the ambition of a free flow of data in Europe.

The data revolution is not just about opening up data, it is about setting up framework conditions for data to be accessed easily and re-used. This means establishing policies that are global enough to address the diversity of data providers, data formats and tools. At the same time, these policies should be specific enough in order to be implemented in practice and avoid data silos from being created. Or, worse, sectors from opting out which would lead to a fragmented open data market and undermine the extent of the expected benefits for data users and society as a whole.

As underlined in the Open Data Maturity in Europe assessment conducted for 2017,[1] countries are at the tipping point and are invited to step up the game to pursue their efforts in opening up their data. More consistent data and metadata quality is expected as well as a more coherent approach to interoperability and data infrastructures.

The present book addresses these two aspects and highlights the importance of policy, evaluation as well as topics that are often underestimated such as interoperability and data infrastructures. It provides guidelines as well as illustrations derived from international best practices.

All actors are uniquely positioned to take part in the data revolution. It is all about getting started and opening your world to open data.

Wendy Carrara
Principal Consultant, Capgemini
Manager of the European Data Portal
Paris, France

[1] European Commission, Open Data Maturity in Europe 2017, November 2017, https://www.europeandataportal.eu/sites/default/files/edp_landscaping_insight_report_n3_2017.pdf

Preface

Motivation

The public sector is information-rich by nature. The opening of data by public organizations is a recent phenomenon in which public sector information is made available and thus can be combined with other data sources and used by citizens for a variety of purposes, including improving the public sector, inspiring business innovation and establishing transparency.

As data can often be generated and provided in huge amounts and through multiple sources, specific needs for processing, curation, linking, and visualization result in the need for open data approaches. Pipelines in the forms of APIs are being created, in which open data is transmitted in real time, for creating new applications and changing citizen behaviour. Cloud services are in parallel changing the ways of providing and using open data, based on vast virtualized resources offering security, privacy and scalability. Data analytics fill in the decision-making process, for citizens, businesses and administrations, providing new ways to model, simulate and even co-create the future.

Although the opening and use of data offers huge potential, how this potential should be exploited is not yet clearly understood. All these developments impact the operation of governments, their relationship with private sector enterprises and the society. Changes at the technical, organizational, managerial and political level are needed, impacting the needed capabilities, policy-making and traditional institutional structures.

This book is inspired by the many open data developments that currently take place, including the following:

- Society has become more data-driven, and more and more data is becoming available from a large variety of sources and actors. This data is often fragmented and provided in different forms. The data can be used under different conditions, and many barriers still exist for the use of open data.
- Over the last decade, various projects have started to address open data challenges and to stimulate the open data movement. These projects are powered and

supported by the European Commission, policy makers, researchers, ICT vendors and the citizens that are actively involved through various open data infrastructures. The ongoing projects focus on different challenges, but oversight is often lacking.
- Many national governments, but also local governments, are developing open data policies and open data portals. There is no 'best policy' or 'best portal', and policies and portals are context dependent.

Aimed Contribution

This book aims at presenting the latest research findings such as theoretical foundations, principles, methodologies, architectures and technical frameworks based on solid and successful cases and lessons learnt from the domain of open data.

The book will contribute to the systematic analysis and publication of cutting-edge methods, tools and approaches for assisting the relevant stakeholders in their quest for more efficient data sharing policies, practice and further research. The topics of the book include (but are not limited to):

- An introduction to open data concepts and definitions, e.g. open data benefits, societal challenges, perspectives on open data and stakeholders
- The open data landscape, e.g. historical developments and an overview of important open data portals and projects
- The open data life cycle, including steps that organizations take in opening data and steps that users take, and the steps for creating benefits and public value with open data
- Open data policies, e.g. the European Public Sector Information Directive, the US open data policy, the Open Government Partnership and national open data policies
- Organizational issues, e.g. administrative processes and activities, organizational risks and potential negative effects
- Interoperability, e.g. interoperability building blocks, metadata and Linked Open Data
- Technologies, e.g. infrastructures, architectures and visualizations
- Business models, e.g. data use outside the government, strategies for making money with open data, and citizen science
- Evaluation, e.g. open data portal evaluation and open data benchmarks
- Research directions, best practices and guidelines

Organization

The book chapters are written from three different perspectives: the open data publisher/public servant perspective, the entrepreneurial/developer perspective and the researcher/journalist perspective. The book is organized along nine chapters, from initial concepts to policies, processes, systems and impact, business potential and research future. The chapters are as following:

Chapter 1: The Open Data Landscape

Creating value by opening and using data is the ambition for many governments. The open data landscape consists of a many, interacting stakeholders that are using all kinds of software to process data. The stakeholders play different roles and their engagements are necessary. Often by combining various dataset, value is created. The objective of the opening of data ranges from transparency, accountability to stimulate innovation by firms. The global landscape shows that countries take various approaches and are in various stages of development. Various instruments are available to measure and benchmark open data efforts. There is no single recipe to create value from data. Some apps are successful, whereas most data is not used. Opening of data might come at a risk. Privacy or sensitive data might be opened or incurred conclusions might be drawn from data. Measures to reap the benefits of open data and avoid the dark site are needed. Finally, recent developments are sketched which shape the open data landscape.

Chapter 2: The Multiple Life Cycles of Open Data Creation and Use

Since the process of open data publication affects the re-use of them and hence the generation of value from them, in this chapter we are going to identify the major step towards publication and usage, analysing different scenarios from the publisher's side. After discussing the publication procedure, we are going to identify the outer cycle of use and re-use analysing usage scenarios about different kinds of data (linked or big) as well as scenarios in different contexts: the researcher's and the pro-sumer's views. This chapter will also present an extended open data life cycle regarding the publication plan resulting the two levels of the cycle: (a) steps towards publication of open data ensuring transparency-by-design (open licence, etc.), quality-by-design (metadata, data structures, timeliness, etc.) and the appropriate functionality (type of data, APIs, user collaboration and feedback, data analysis and visualization) and (b) steps towards exploitation, value generation and re-use. The communication and feedback steps of the cycle and its associated social media mechanisms (Web 2.0 functionality) are the ones that close the feedback loop. Finally, three principles for open data have been identified and presented.

Chapter 3: Open Data Directives and Policies

In developing open data policies, organizations aim to stimulate and guide the publication and use of data and to gain advantages from this. Often open data policies are guided by a high-level directive, such as those of the United States and the European Commission. Currently, a multiplicity of open data policies is under development at governmental agencies at various administrative levels. In this chapter, we explore the elements and characteristics of open data directives and policies. We provide examples of elements of directives and policies, we discuss existing open data directives and policies, we provide an example of the elements of the Dutch national open data policy and we discuss lessons learned from open data policy development. This chapter shows that several frameworks for comparing open data policies have already been developed, and they show that a wide variety of open data policies exist. Existing policies have a different focus and open data policies may encompass different elements. The elements of open data policies that we describe in this chapter are not covered by every policy. There is variety in the policy environment and context, the policy content (the input), the performance indicators (the output), the attained public values (the impact) and policy change or termination (the feedback). The differences between open data policies may indicate that open data policies stimulate the provision and use of open data in different ways, and this could reveal opportunities for learning from each other.

Chapter 4: Organizational Issues: How to Open up Government Data?

Governments create and collect enormous numbers of datasets, for instance concerning voting results, transport, energy, education and employment. These datasets are often stored in an archive that is not accessible for others than the organization's employees. To attain benefits such as transparency, engagement and innovation, many governmental organizations are now also willing to give public access to this data. However, in opening up and in publishing their data, these organizations face many issues, including the lack of standard procedures, the threat of privacy violations when releasing data, accidentally releasing policy sensitive data, the risk of data misuse and problems with data ownership. Opening up governmental data requires various changes at different organizational layers. These issues hinder the easy publication of government data. In this chapter we first discuss issues that governmental organizations face when opening up their data. We give an overview of all the issues and then discuss each of them in detail with a related example from the open government domain. Subsequently, we provide guidelines for governmental organizations that want to open up their data. Such guidelines can be used by public organizations to improve their open data publishing processes. Ultimately, the implementation of the guidelines reduces barriers, stimulates the publication of government data and contributes to attaining the benefits of open data. Discussions with practitioners showed that the principles could improve the open data publication process.

Chapter 5: Open Data Interoperability

Interoperability is of utmost importance when it comes to exchange of data between different entities or organizations, in particular in a cross-border scenario. Related interoperability aspects already start at the very low end of the data stack, namely regarding used file formats, interfaces of data platforms, as well as defined exchange protocols. But even with these key aspects in place, the usefulness of exchanged data can be diminished by low metadata quality. Therefore, the assessment and semi-automated improvement of metadata is a key for successfully establishing open data interoperability. Yet, in some use cases, these basic conditions are not enough. The exchange of data can be tricky, especially in cases across domains or across borders, introducing another barrier in the form of domain-specific language or different national languages in general. To overcome these semantic gaps, ontologies and the possibility to link data for an improved understanding of a given context are imperative. Therefore, this chapter sheds light on state-of-the-art approaches in the domains of interoperability building blocks, metadata quality, as well as ontologies and linked data in the field of open data.

Chapter 6: Open Data Infrastructures

Data represents a key asset in virtually any aspect of society and economy and therefore triggers a radical shift of importance of the establishment of data infrastructures. Associated to this shift is the necessity of these infrastructures to feature a high level of resilience, robustness, as well as the required scalability. Yet, access to open data comes not only in the form of a solid infrastructure, understanding the interaction of the data and the stakeholders using it are at least as important. Examples can be found in the domain of open science and open research, enabling citizens to engage in the ongoing development and usage of open data as well as in the domain of e-participation. While all technological facets are important, trust and transparency may not be neglected to ensure the sustainability of an envisioned open data infrastructure. The chapter will therefore provide details regarding functional requirements as well as a layer of trust via the use of blockchain technology towards the realization of public sector applications. Finally, the chapter also introduces two pilot projects regarding open data infrastructures in Austria and Germany.

Chapter 7: Open Data Value and Business Models

The chapter looks into the process of turning data released in an open format into meaningful and valuable innovations both by the public and the private sector. More specifically, the discussion focuses on how such innovation may be enacted. Starting from a definition of the open data value chain the chapter subsequently shifts the focus towards understanding which business models may be leveraged. Finally, a number of real-life use cases are discussed to exemplify the concepts presented. On the one hand, such processes represent a great opportunity for private and public organizations while, on the other, they pose a number of challenges having to do with creating the technical, legal and procedural preconditions as well as identifying appropriate business models that may guarantee the long-term financial viability of such activities. As a matter of fact, while information sharing is widely recognized as a value multiplier, the release of information in an open data format through creative common licences generates information-based common goods characterized by non-rivalry and non-excludability in fruition, an aspect posing significant challenges for the pursuit of sustainable competitive advantages. The objective of the chapter is to shed light on some of the challenges highlighted above, with particular reference to the business models that may be adopted for igniting data-driven value generation activities. More specifically, the chapter starts by providing some background on a few key concepts having to do with the notion of value, the economics of information, business models and the open innovation paradigm. Subsequently, an overview of the most prominent studies on business models for open data is presented. Finally, the main exploitation opportunities and some real-life cases will be discussed to exemplify a number of good practices of open data valorization in both the private and the public sector.

Chapter 8: Open Data Evaluation Models: Theory and Practice

Different models and procedures have been used for the evaluation of open data and their portals examining different aspects of them. In this chapter we are going to identify the subjective and objective measures for the evaluation of open data as well as the platforms offering them. Indicators for the measurement of impact achieved in the form of open data benchmarks will be analysed and proposed for each case of the life cycle. Furthermore, an analysis of the current assessment models is presented with pros and cons in each case. This chapter will present and analyse the existing evaluation models in the information systems domain. It will also showcase different aspects of evaluation through application examples. A taxonomy of measures and metrics was created towards the evaluation of quality of open data, their portals and their functionalities. Finally, guidance for constructing an evaluation framework is provided incorporating different evaluation aspects.

Chapter 9: Open Government Data: Areas and Directions for Research

The chapter aims at illustrating the present and oncoming research domains around open data deployment, curation and use. Open data has been a thriving multidisciplinary research domain, gathering researchers and practitioners from various disciplines like information systems, databases, process management, social sciences and law. Although systems, approaches and literature on open data have been evolving, together with research performed in various projects and initiatives worldwide, a systematic analysis of the research areas around open data is still missing. In this chapter, the taxonomy of research areas in the open data domain is presented, stemming from a thorough state of the art analysis and deliberation with experts at an international scale. The taxonomy contains organizational, technical, semantic and legal issues that need to be researched in the coming years, organized in several layers. For each of the more than 50 nodes/research areas, the basic literature is presented and the main targets for researchers over the next years are analysed. The chapter also discusses multidisciplinarity issues on open data and gives an overall view of how research on open data can assist societies in tackling important societal problems. Conclusions give the reader the possibility to understand the key barriers to overcome and the most important research gaps to fill, in order to have successful open data implementations under different deployment scenarios.

Four appendices are adding useful resources for the reader, the researcher and the practitioner of open data with References, Abbreviations, Terms Index and Authors Biographies.

As a Conclusion

Today, that this book is made available to its readers, the open, big and linked data community is considered a significant factor that can help tackling the economic, political and organizational challenges our societies face.

Luckily, infrastructures and practices like big data management and processing, cloud computing, internet of services, and things, electronic participation, social media, policy modelling, simulation and the new evolutions in mobility, interactivity and collaborative nature of software and human actors have the collective potential of altering our world for the better.

It seems though, that this better world will only appear if these resources and technologies do not stay under the control of the few, but and are provided openly, usually with none or minimum cost to citizens, communities and certain forms of enterprises. It is only through open data and open services, under inclusive regulation and a vision for creative destruction, that societies can entertain significant gains from computers, devices, networks and their software.

May the concepts, methods, tools and experiences presented in this book serve as your useful companions, in this quest for a better world.

Samos, Greece — Yannis Charalabidis
Delft, The Netherlands — Anneke Zuiderwijk
Samos, Greece — Charalampos Alexopoulos
Delft, The Netherlands — Marijn Janssen
Krems an der Donau, Austria — Thomas Lampoltshammer
Turin, Italy — Enrico Ferro
July 2018

Acknowledgements

This book is the result of the collective work of, primarily, the authors. But it is also a product of openness and collaboration with more than one hundred other scientists, industry experts and practitioners in the fields of open, big and linked data. We are highly grateful to all of them involved in overall guidance, stimulation of the community, the review process and the book finalization.

Many thanks go to colleagues from the ENGAGE e-Infrastructures visionary project, where together with our friends from the National Technical University of Athens, IBM research Haifa, Microsoft Innovation Centre Athens, EUROcris, Science and Technology Facility Council, Fraunhofer FOCUS, Intrasoft International and so many more projects and organizations, we discovered, we tried and we learned.

We would also like to thank Daniel Santiago, Timos Sellis and Wendy Carrara for their warm forewords in this book.

Special thanks also go to the publisher's team and particularly to Kelly Daugherty for her professional guidance, support and feedback – decisive for keeping this project on time and quality.

Finally, a big hug for our family members and close collaborators, for their love and support.

This book is devoted to Lefki, Patrick, Penny, Henri, Daphne, Karin, Katrin and Giulia.

The Authors

Contents

Chapter 1
The Open Data Landscape

"Open data has many different aspects: objectives and benefits for a variety of stakeholders, but open data also has a dark side."

1.1 Creating a New World of Open Data

The opening of data has grown tremendously over the past decade. More and more datasets have been opened to the public, application programming interfaces (APIs) gave been design for enabling the public to make use of real-time data and new apps based on this data have been developed. Data about policy-making, software code (open sources), documents, minutes, financial data and so on has been opened resulting in a large repository of government data that can be on open data portals and government websites. Nevertheless the potential is even higher, as most of the data is still closed and is not directly accessible to the public. Furthermore, more and more data is collected and can be share in nowadays words driven by The Internet of Things (IoT). The IoT consist of devices that are able to collect data such as GPS (geographical location), compass, temperature, movement, pollution and so on. Devices collecting data combined with data analytics are expected to transform the government and society. This can provide insight into the energy consumption of smart cities (https://amsterdamsmartcity.com/projects/energy-atlas) or the pollution (http://airindex.eea.europa.eu/). These initiatives are all driven by the opening of data and extended by user-friendly apps to enable large use by the public.

Over the course of the past few decades, many governments have imitated all kinds of projects to open their data to the public. This practice have been followed by private organizations that started also opening some of their data resulting in the creation of business value (Zuiderwijk, Janssen, Poulis, & van de Kaa, 2015). The availability of open government data has grown significantly, with pressure being placed on all kinds of public organizations to release their raw data for the good.

The movement of opening data resembles a move from a closed to an open system (Janssen, Charalabidis, & Zuiderwijk, 2012). Open systems are encountered with uncertainties from the environment and are less predictable and therefore not

Y. Charalabidis et al., *The World of Open Data*, Public Administration and Information Technology 28, https://doi.org/10.1007/978-3-319-90850-2_1

easy to manage. By opening some data, also insight into the functioning of the government is revealed. This might be viewed as a risk by some public servants, whereas others views this as a way to strengthen the democratic system by creating transparency and accountability. The public is empowered by giving it the data and the means for making sense out of the data. Also businesses can benefit from the opening of data and enrich their existing products or develop new products (Zuiderwijk, Janssen, Van de Kaa, & Poulis, 2016).

Open data and open government are related. Open governments objectives relate to creating transparency, accountability and engagement to strengthen the governance and empower citizens. The opening of data is a means for this, but not sufficient as also institutional measures might be necessary. This includes steps to take measures when corruption or fraud is detected using open data. Open can include Open Government Data (OGD), but also Open Business Data (OBD) or Citizen-Generated Data (CGD). The latter is data collected by citizens, which can be done by using IoT devices.

Also the public can become part of the policy-making processes. Ordinary people can become part of the policy-making and might collect data, process data, combine it with other sources to create new insight to help policy-makers. In this way, new opportunities for involving the public in policy-making processes become available. Also citizens might process data, enrich data, combine it with other sources and might even collect their own data (for example through the use of their mobile phones).

Open data can be looked at in various ways and there are various definitions available. Instead of giving another formal definition we prefer to look at the characteristics of what makes data really open. The Sebastopol principles elaborate on what makes data "open data" (Malamud et al., 2013). Open data should be primary data, published in a timely manner and allowing diverse groups with different interests to take advantage of this. This includes the following aspects.

- Data must be complete
- Data must be primary
- Data must be timely
- Data must be accessible
- Data must be machine processable and made online in persistent archives
- Access must be non-discriminatory
- Data formats must be non-proprietary
- Data license must be unrestricted and bear no usage costs
- Also data should be as accurate as possible.

Indeed most of the data will not meet this list of requirements. Nevertheless, data is only truly open if most of these criteria are met. In this book the 5 stars model of Tim Berners-Lee will be discussed which provides insight into the maturity of the data, where each additional star means that the data meets the criteria of the previous steps (http://5stardata.info/en/).

1.2 Historical Developments

The opening of data by the government has already a long history. Traditionally data was only opened upon request by the public. The right to have access to data is central to the Freedom of Information (FOI) Act. Although many countries already had a FOI act before, the FOI is linked to the article 19 of the 1948 Universal declaration of basic human rights of freedom of expression (http://www.un.org/en/universal-declaration-human-rights/). Many countries have Freedom of Information Acts (FOIAs) in place in which citizens can ask for information (Petticrew & Roberts, 2008). FOIA allows the public to ask for (partial) disclosure of information and data is not released yet. The amount of FOI varies overtime and the requests are often coming from the same stakeholders who have the opportunity and time to ask for this data. Governments have developed procedures and processes to receive FOI requests, process them and give answers. Some people have misused this act to ask many questions requiring many resources of the government. Yet, the asking for information cannot be used by companies for innovating their products or developing new value propositions. Also following FOI is a cumbersome and sometimes lengthy procedure which makes it less suitable for certain applications.

Whereas, FOI is based on the 'based upon request' principle, the proactive provision of data to the public is based on the 'open by default' principle. The pro-active opening of data streams is initiated by Obama's Memorandum on 'Transparency and Open Government' published in 2009 (McDermott, 2010). Obama's Memorandum encourage the active disclosure of public data, instead of waiting for requests. This Memorandum resulted in the development of open data portals (see for example www.opendata.gov) in which open data is released to the public. Policies stimulating the opening of data were developed and public organizations were asked to start with the release of their datasets. The USA example served as a sources of inspiration for many other governments. For example, the EU Public Sector Information (PSI) directive, which is focused on making public sector data available and ensuring a level playing field (European_Parliament_and_Council, 2003).

The Open Government Partnership (OGP) is a partnership launched in 2011 to stimulate open government by empowering citizens, fighting corruption, and harness new technologies to strengthen governance (https://www.opengovpartnership.org/). The opening of data is an important means for this. Opening up government data is a voluntary initiative that countries can join and is aimed at securing and taking actions to strengthen governance.

1.3 Objectives of Open Data

The objectives of open data relate to coming closer to an open government, stimulating and enabling private sector innovation, and stimulating engagement and participation of stakeholders like citizens and companies. The three areas are visualized in the figure below. Government should become transparent and accountable by

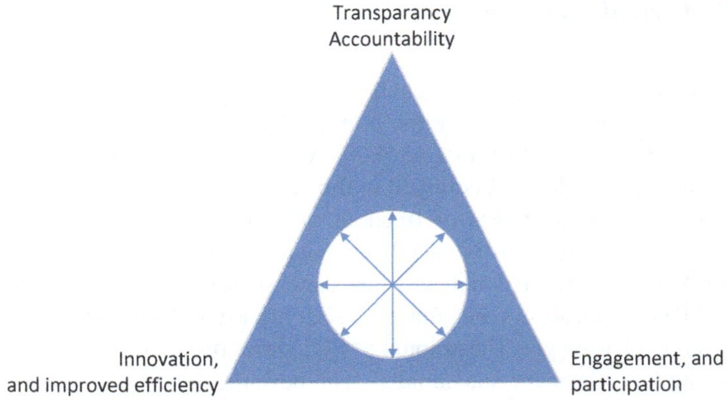

Fig. 1.1 Objectives of open government data

promoting the public right of access to information (McDermott, 2010). This can even be viewed as a requirement of a democratic system and concerns the opening of data about the functioning of the government and their decision-making.

The second one is has economic motives to encourage the opening of government data which can be used by companies and society to create value. The government has a lot of data that, when opened, can be used to create new entrepreneurial activities, to add value to existing services offerings, or to create new insights which enable to improve business (Fig. 1.1).

The third area of open data objectives concerns the stimulation of engagement and participation. Open government data gives governments a new means to communicate their activities to citizens and other stakeholders and to invite various actors to give feedback on government activities and participate in them.

1.4 The Stakeholder Landscape

There are often many stakeholders involved in the opening of data. Often the actor that is sharing the information is not necessarily the organization that collected the data or processed the data. Many more organizations and departments might be involved. Some organizations might support the opening of data like software vendors, whereas other stakeholders are directly involved. The stakeholder landscape adds to the complexity of open data as responsibilities for opening data might not be clear, the ownership of data cannot be defined easily and many parties should collaborate for opening data (Table 1.1).

1.5 Open Data and Big Data: A World Apart?

The field of open data consist of the many areas referred to the term 'data' in general as shown in the figure below. The origin of the data can be the government, businesses or citizens. Open data refers to the situation that data is made available

Table 1.1 Overview of main stakeholders

Stakeholder name	Simplified Stakeholder descriptions
Politicians	Often in charge of creating open data policies and assigning resources and budget for realizing open data policies.
Data collector	These organizations are collecting data. These are often public organizations, but can also be citizens who collected data or businesses.
Data processor	Data needs to be analyzed and processed. Many organizations have data analysts who are able to make sense of data.
Data publishers	The organizations that publish the data. Often in open data portals, but also using APIs and other means.
Infrastructure providers	Data needs to be communicated over a secure and reliable network infrastructure.
Companies and businesses	Business might use open data to add value to their existing products and services or analyze data for their own advantages. They might even use open data for empowering their lobbyist.
Infomediaries	Infomediares (information + intermediaries) are organizations that collect and process open data. They have more resources and can lower the threshold for citizens to use the data
Citizens	Citizens are often the intended users of data. The idea is that they will engage with the government, use open data to scrutinize decision-making in the government to create transparency and held them accountable. Ultimately this should result in a higher trust in the government, however, this is challenged by many researchers.
Software vendors	Often many software providers are involved to collect, process or to publish data. There are specialized organizations that are able to anonymize documents for publications. There is software that can be used to analyzed and visualize data, and there is separate software to make data available.
Regulators and ombudsman	Regulators have two different roles on the one hand can regulators be in charge of ensuring compliance with the data regulations and data protection act. On the other hand they can use the data for investigating governments and to look at what is happening.

outside the own organization for use by others. Ideally to everybody without any restrictions for further use. Yet, licenses might limit what can be done with the data. Often data might not be used for commercial use which limits the use for businesses to make profit from data.

Big data is commonly characterized by several Vs, including Volume, Velocity, Variety (McAfee & Brynjolfsson, 2012). Gandomi and Haider (2015) add another three Vs to this list; Value, Variability and Veracity. The essence of big data is that this concerns data that cannot be handled in traditional ways (Elgendy & Elragal, 2014a). Big data is closely related to Big Data Analytics (BDA) which are needed to create value of the data (Elgendy & Elragal, 2014a; Holsapple, Lee-Post, & Pakath, 2014). Although big data and open data are closely related, yet they are not the same as, big data is characterized by its size and open data by its availability (Janssen, Matheus, & Zuiderwijk, 2015).

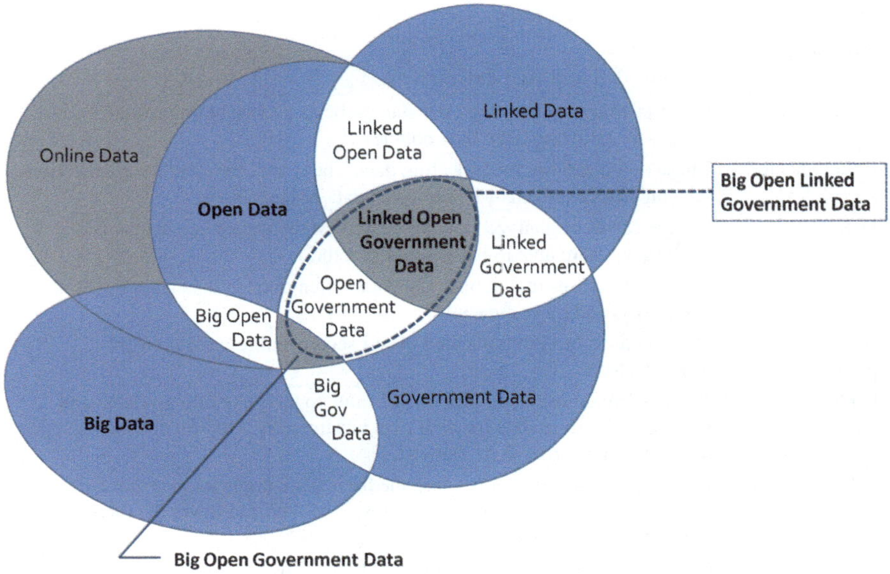

Fig. 1.2 Overview of the field of open data

Data often originates from many sources which are often beyond the control of a single actor like social media and devices. Therefore there is a need to link data to created 'linked data'. Linked data is about relating structured data into machine-readable format that can be semantically queried (Bizer, Heath, & Berners-Lee, 2009). This enables the searching for the data, but also to combine different datasets to create value from them. The creation of value from data requires combining large datasets originating from different and heterogeneous data sources (Janssen, Estevez, & Janowski, 2014). *Big Open and Linked Data* (BOLD) is an acronym often used for depicting to the use of data in the digital age referring to the changing nature of data (Janssen et al., 2015) (Fig. 1.2).

1.6 Benefits of Open Data

There are many benefits can be accomplished with the opening of data that range from political to technical benefits (Janssen et al., 2012) as listed in the table below. The benefits are not mutually exclusive, but they are a good starting point for making the case for opening data (Table 1.2).

Table 1.2 Overview of benefits of open data

Category	Benefits	Description
Political and democratic	More transparency	The creation of more insight into the functioning of the government
	Democratic accountability towards citizens (users)	The answering for the actions taken to the public. The ability for other organizations to scrutinize the government ad to check their actions
	Trust in government	The creation of more trust in the government by acting in a transparency and accountable way.
	More participation and self-empowerment public engagement	The ability of citizens and business to participate better in democratic decision-making and have a deeper understanding of the issues.
	Empowering the public	Providing the means to understand and participate in decision-making.
	Equal access to data	Leveling the play field by ensuring that the public have the same data as policy-makers have.
	New governmental services for citizens/improvement of citizen services	The creation of new data-driven services that can be more customer-centric and deal with societal problems.
	Improvement of policy-making processes	By providing information policy-makers can tap into the wisdom of the crowds and in this way improve the policies.
Organizational	More visibility for the data provider	Organizations might be viewed as innovative and gain more visibility
	Improvement of citizen satisfaction	The opening of data can result in more satisfied citizens.
	Use of the wisdom of the crowds:	By opening of the data organizations can tap into the intelligence of the collective
Innovation	Stimulation of knowledge developments	The opening of data can result in new innovative applications with the data by others, which in turn can stimulate innovation within the government.
	Creation of new insights in the public sector	Opening of data can create new insights. Also government can start to use each other data and create new insights.
	New (innovative) social services	New services can be developed for the public
Economic	Stimulation of competitiveness and innovation	Data can be used to create new businesses, for the development of new products and services or to extend current service offering
	Economic growth	The creation of a data economy. Data is fueling economic growth. Creation of a new sector adding value to the economy
	Availability of information for investors and companies	Companies can use open data to determine where to invest and where to locate their businesses.

(continued)

Table 1.2 (continued)

Category	Benefits	Description
Operational and technical	Reuse of data	The ability to reuse data / not having to collect the same data again and counteracting unnecessary duplication and associated costs (also by other public institutions)
	Improve administrative processes and policies.	The opening of data and feedback gained can be used to optimize administrative processes and policies.
	Improving the quality of data	External quality checks of data (validation) and the public can help to improve the quality of data.
	New data	The ability to merge, integrate and mesh public and private data. Creation of new data based on combining data.

Based on Janssen, Charalabidis, and Zuiderwijk (2012)

1.7 The Dark Side of Open Data

All too often the focus of politicians is on the benefits and the possibilities of open data, whereas the public administration is afraid of the risks of opening data. The opening of data might require considerable resources, however, the opening might not result in any public value at all. Resources might be wasted on releasing data that are not used or even not relevant. Zuiderwijk and Janssen (2014a, 2014b) found the following issues that might hinder the opening of data, although there are many mechanism that can be used to overcome them. For example, privacy-enhancement mechanisms (PEM) are often used to comply with the data protection act (Table 1.3).

The risks might result in inertia and the avoidance of the opening of data. Nevertheless most of the issues can be dealt with, however, the costs needed to deal with them often hinder the opening of data. Budgets are tight and many organizations have no or limited budget for opening data.

1.8 Developments

Whereas much focus is still on opening data, there are developments to have 'openness by default' and "transparency-by-design". These concepts refer to the situation where software is designed in such a way that when data is collected the data is collected in such a way that the opening of data is possible (Janssen, Matheus, Longo, & Weerakkody, 2017).

Data is fragmented, described in different formats by different organizations. In many portals data is opened, but not well-described which makes searching for data and the interpretation of the usefulness of datasets difficult. Semantic descriptions,

Table 1.3 Overview of risks of open data

Category	Risk	Description
Legislation	Non-compliant	All kinds of legislation might be applicable from different domains. There might be unawareness of which legislation might be applicable.
	Privacy	The data protection act poses strict requirement on what can be published and what cannot be published. Although there are PEM, the privacy of persons can be violated unintentionally or data might be used for another purpose than what it was collected for.
Governance	Responsibilities	Difficulties with data ownership and stewardship. Unclear responsibility and accountability.
	Maintaining quality	Unclear if data is updated and what the data quality is.
Interpretation	Bias	Published data can be biased
	Ambiguity misinterpretation	Misinterpretation and misuse
Data quality	Poor data quality	Data might have different qualities (completeness, accuracy, timeliness). Decisions can be made on poor information quality or the wrong insights can be created.
	Timeliness	The most recent data might not be available. Embargo period prohibits the publication of recent data

Based on Zuiderwijk and Janssen (2014a; 2014b)

adding metadata and linking the data improves the use of the data. In addition, meta-search engines, have become available which have indexed many data portals. Also there are data standardization working groups that are developing comprehensive meta-data models for describing open data like CERIF (Jeffery, Houssos, Jörg, & Asserson, 2014).

Also automatic annotation and retrieval software has been developed. Data range from structured to unstructured data and data might not be used easily. Unstructured data can be transferred into structured data by annotating the data. For example, this happens when somebody adds the persons in a picture on Facebook. More and more automatic tools can be used to automatically annotate unstructured information.

Also in the field of -statistical data and visualization there are initiatives to make the collection, linking and analysis of Linked Open Statistical Data (LOSD) easier (Kalampokis, Tambouris, & Tarabanis, 2017). In the ideal situation no knowledge of software is needed and by drag and drop applications statistical data can be combined and visualized.

Chapter 2
The Multiple Life Cycles of Open Data Creation and Use

Open data can be defined as data that is free of charge or provided at marginal cost, under an open licence, machine readable, and provided in an open format

2.1 Introduction

Different terminologies have been suggested towards the description of various models of open data. The open data life cycle, the open data value chain or the open data process (Zuiderwijk, Janssen, Choenni, Meijer, & Alibaks, 2012) are terminologies illustrating different purposes – practical guidance or analytical understanding – and foci. Whereas value chain models – that will be further analysed in Chap. 7 – focus more on the creation of value during open data usage, the life cycle models aim to structure the handling of the data itself. Existing process models focus on activities within public administrations, such as generating (create/gather), editing (pre-process and curate) and publishing the data without paying too much attention on the outside-use and re-use processes.

In order to fully exploit the benefits of open data, traditional "one-way street" open data practices and initiatives should be replaced by an open data ecosystem, i.e. an approach to open data that focuses not only on data accessibility, but also on the larger environment for open data use—its "ecosystem" (Pollock, 2011; World Bank Group, 2015). An open data ecosystem can be defined as a cyclical, sustainable, demand-driven and environment-oriented around agents that are mutually interdependent in the creation and delivery of value from open data (Boley & Chang, 2007; Harrison, Pardo, & Cook, 2012; Heimstädt, Saunderson, & Heath, 2014).

Because of these many interdependencies, open data ecosystems should be studied as a whole, by investigating both the user and the publisher sides of the life cycle as well as the relation to each other. (Susha, Janssen, & Verhulst, 2017) in their proposal for a user-centric and interdisciplinary research agenda to advance open data: *"To realize its potential there is a need for more evidence on the full life cycle*

Y. Charalabidis et al., *The World of Open Data*, Public Administration and Information Technology 28, https://doi.org/10.1007/978-3-319-90850-2_2

of open data – within and across settings and sectors". In other terms, interdisciplinary open data research should investigate the open data life cycle in all its phases and address open data developments in different domains.

The open data life cycle is a conceptualization of the process and practices around handling data, starting from its creation, through the provision of open data to its use by various parties. In addition, the characteristics and interests of different stakeholders involved are hardly recognized and taken into account. Analysing different data life cycle models from technological (data curation, big data and linked data) and stakeholders (publishers and users) perspectives, this chapter introduces an advanced open data life cycle model based on all the above identifying associated tools for each stage of the cycle, as well as, the transitions and interdependencies between different phases.

Moreover, the advent of Linked and Big Data as well as the collaboration capabilities of Web 2.0 paradigm reformed the landscape of open data since they introduced enhanced capabilities. These advanced capabilities, in their turn, introduced different concepts, solutions and complexity in the data re-use, storing, analysis, and publication processes.

This chapter introduces the new requirements for open data provision and usage in terms of different technologies (linked and big data) along with the accompanying impediments as well as an overview of the existing life cycle models for open data in Sect. 2.2. Section 2.3 presents an accumulative model derived from the conjunction of the two different stakeholder sides as well as the duality of the users' roles in an open data ecosystem. It also defines different tools and methods in each step of the open data life cycle concerning the requirements of different types of data. Section 2.4 familiarizes different uses of the open data life cycle presenting the open data life cycle from the perspectives of the two different stakeholders, namely, the open data producer and the open data user. It also describes the application of the open data life cycle model in the research domain supporting the development of a Scientific Data Infrastructure (SDI). Finally, Sect. 2.5 concludes the chapter referring to the principles underpinning the life cycle and the open data ecosystem.

2.2 New Requirements for Open Data Provision and Usage

2.2.1 Linked Data

The linked data paradigm puts an emphasis on the structure of the data using triples and description based on RDF (Resource Description Framework) vocabularies as well as in storing technologies (SPARQL) solving also the issues of uniqueness and metadata. Linked data is a method of publishing structured data so that it can be interlinked and become more useful through semantic queries. The concept builds upon standard Web technologies such as HTTP, RDF and URIs, but rather than using them to serve web pages for human readers, it extends them to share information in a way that can be read automatically by computers. This enables data from different sources to be connected and queried (Soylu, Mödritscher, & De Causmaecker, 2012).

When we are dealing with linked data and since it is a quite novel technology, there are some important impediments that should be taken into account (Auer et al., 2012). First of all, linked data uses RDF Data Management Systems (i.e. SPARQL) which are more challenging than the relational data management. Ways of limiting this performance gap include column-storage technology, dynamic query optimization and other. Secondly, creating and maintaining links in a (semi-) automated fashion is still a major challenge and crucial for establishing coherence and facilitating data integration. New linking approaches should yield high precision and recall, which configure themselves automatically or with end-user feedback. Thirdly, since linked Data on the Web is mainly raw instance data, data integration, fusion, search and many other capabilities need to be linked and integrated with upper level ontologies. Fourthly, the quality of content on the Data Web varies, as the quality of content on the document web varies. Finally, since Data on the Web is dynamic, it is essential to facilitate the evolution of data while keeping things stable in methods development to spot problems in knowledge bases and to automatically suggest repair strategies. An example of linked data usage is presented in Sect. 2.4.4.

2.2.2 Big Data

The potential benefits of Big Data are significant, but many technical challenges should be addressed to fully accomplish those benefits (Jagadish et al., 2014). One of the most renowned challenges is the sheer size of the data. However, there are others such as Variety and Velocity completing the 3 V's of big data. Variety refers to heterogeneity of data types (structured and unstructured) originated by disperse data sources aiming at data representation and semantic interpretation. Velocity implies the time frame the data should be analyzed according to the rate of data arrival. Further important requirements have been detected since big data applications began such as veracity (reliability), variability (complexity) (Gandomi & Haider, 2015), privacy and usability (Jagadish et al., 2014).

Dealing with big data is a quite exhaustive task bringing in changes in technological and analytical level of data processing as well as in data storage with the most prominent technology to be the NoSQL databases. The advent of big data alternates the importance of the life cycle steps placing more focus on the "create", "process" and "store" steps of the life cycle. Technologies for covering these steps are the major concern at the moment. New analysis methods (indexing algorithms towards timely data analysis) have derived and applied on big data. An example of big data usage is presented in Sect. 2.4.2.

2.2.3 Web 2.0

In addition following the Web 2.0 paradigm (Alexopoulos, Loukis, & Charalabidis, 2014; Charalabidis, Alexopoulos, & Loukis, 2016) there is a new generation of OGD platforms and virtual environments trying to fill the gap of communication between data users and data providers through closing the feedback loop as well as creating the notion of data 'pro-sumers'. This shifts the paradigm towards highly active users, who assess the quality of the data they consume and mention weaknesses of them and new needs they have; who often become both consumers and providers of data is characterised by advanced capabilities to data users for commenting, rating, processing in order to improve them, adapt them to their specialized needs, or link them to other datasets (public or private); and then uploading-publishing new versions of them, or even their own new datasets. This systemic view of open data could be used to the development of new solutions matching supply and demand and utilising the innovation aspect of open data.

Zuiderwijk, Loukis, Alexopoulos, Janssen, and Jeffery (2014) proposed an open data electronic marketplace with enhanced capabilities for both producers and users. The new marketplace also supports the data pro-cumer enabling advanced publication procedures connected with the appropriate tools. The EU-FP7-ENGAGE project could be seen as such a marketplace, since its functionality supports all the identified requirements except the payment and value definition procedures which have not been realised in the ENGAGE context. Without the value definition and payment procedures the ENGAGE platform could be seen as a crowdsourcing-based platform for data processing and data exchange among users. The basic and novel functionality of such an architecture is shown in Table 2.1.

Table 2.1 Classical and novel functionality of OGD infrastructures adapted by Zuiderwijk et al. (2014)

Functionality	Stakeholder	Description
Classical open data functionality		
Data Publication	Provider	Support for publication to the providers: tutorials and guiding principles for data uploading
Data Modeling	Provider	Capabilities of flat metadata descriptions (based on a specific metadata models) and data formats
Data Search	User	Simple search via keywords, resource format, publisher, topic categories and countries
Data Visualisation	User	Simple visualisation techniques on specific datasets (maps, charts)
Data Download	User	Data and metadata downloading capabilities. Provision of API.

(continued)

Table 2.1 (continued)

Functionality	Stakeholder	Description
Novel open data functionality		
Grouping and Interaction	Provider/ User	Capabilities for (a) searching for and finding other users/ providers having similar interests in order to have in-formation and knowledge exchange and cooperation, (b) forming groups with other users/providers having similar interests in order to have information and knowledge exchange and cooperation, (c) maintaining datasets/working on datasets within one group, (d) communicating with other users/providers through messages in order to exchange information and knowledge and (e) getting immediately updated about the upload of new versions and enrichments of datasets maintained/worked on within the group, or new relevant items (e.g. publications, visualizations, etc.).
Data Processing	Provider/ User	Capabilities for (a) data enrichment – i.e. adding new elements – fields, (b) for metadata enrichment – i.e. fill in missing fields, (c) for data cleansing – e.g. detecting and correcting ubiquities in a dataset, matching text names to database IDs (keys) etc., (d) converting datasets to another format, (e) submitting various types of items – e.g. visualisations, publications – related to a dataset and (f) datasets combination and Mash-ups.
Data Enhanced Modeling	Provider/ User	Capabilities for description of flat, con-textual and detailed metadata of any metadata/vocabulary model.
Feedback and Collaboration	Provider/ User	Capabilities (a) to communicate own thoughts and ideas on the datasets to the other users and the providers of them through comments, (b) to read interesting thoughts and ideas of other users on the datasets through comments they enter on them, (c) to express our own needs for additional datasets that would be interesting and useful to me, (d) to get informed about the needs of other users for additional datasets and (e) to get informed about datasets extensions and revisions.
Data Quality Rating	User	Rating system against the basic quality aspects of datasets with capabilities to: (a) get informed on the level of quality of the datasets perceived by other users through their ratings and (b) communicate to the other users and the providers the level of quality of the datasets that I perceive.
Data Linking	Provider/ User	Capabilities of data and metadata linking to other ontologies in the web of data (Linked Open Data Cloud). Capabilities of querying data and metadata through SPARQL endpoints.
Data Versions Publication	Provider/ User	Support for publication/upload of new versions of the existing datasets, and connection with previous ones and initial datasets.
Data Visualisation	User	Advanced visualization techniques and visual analytics on specific datasets and/or datasets mashups (maps, charts, plots, series and other)

2.2.4 Models Describing the Data Life Cycle

Most models contain similar elements and differ only regarding semantics, granularity or the extension of the process (Carrara, Fischer, & Steenbergen, 2015). As a first remark emerging from the analysis of Table 2.2, the existence of a perfect life cycle model is not possible based on the various aspects (i.e. curation, preservation) and unique characteristics in each type of data (i.e. linked, big). Different models could be more applicable in different contexts as it can be observed in the examples of Table 2.2.

It is also observed that there are a lot of common stages/steps/phases that could be considered neutral being present in most of the life-cycle models, such as: discovery and acquisition, data organization, publication, integration, analysis, re-use and storage/preservation. These models describe the life-cycle as a sequential, one-dimensional process of activities that an unspecified set of actors repeatedly undertake in order to provide a formerly unexposed amount of data to an abstract general public.

Whereas only making available large volumes of different types of data might result in searching for a needle in a hay stack, the use of predefined views and apps might filter too much information to deliver true transparency. Linked data could be referred as a technology that enables the connection of different datasets in the web of data, in which the searching, acquiring and analysis capabilities are more structured but not too effective. The connection is achieved through the modelling stage of the linked data life-cycle. The modelling stage utilizes vocabularies and generic ontologies (FOAF, SKOS, RDF) for the description of the data in order to establish linkages between different datasets.

Furthermore, these models include only one analytical level. They exclusively take the operational processes of open data publication into account (such as extracting, cleaning, publishing and maintaining data), while largely ignoring the strategic processes (such as policy production, decision-making and administrative enforcement). Thus, the decisions which data will be published, who extracts data, how are data edited, how data can be accessed, which licenses are available, how data privacy and liability issues are treated, and who is involved in these decisions remain underappreciated (Open Data Monitor, 2015).

The data curation model is the only model that could be considered as being comprehensive, since it includes administrative and managerial processes. These more general strategic processes about open data refer to the governance structure, likely to be connected to an organization's ICT and data governance. For example, the planning and the execution of preservation actions throughout the curation life-cycle of the digital material. This would include plans for management and administration of all curation activities in the life-cycle.

The outlined issues point to another blind spot of most open data life-cycle models that these are actor-blind. Until the final model for linked data (section) was conceptualized there were no feedback capabilities and limited capabilities on retrieving, integrate and re-use open data. If at all, institutional characteristics and

Table 2.2 Data life cycle models

Model	Key elements	Part of the open data life cycle covered	Strength(s) of this model	Weakness(es) of this model	Example of how this model can be used
DCC Curation Lifecycle Model	(a) the data itself divided in digital objects and databases, (b) administrative and managerial actions, (c) the basic model and (d) the evaluation actions	Create, Pre-process, Curate, Store, Acquire, Process, Use	Curation preservation of data + Managerial and Administrative Procedures	Ideal model, not very realistic	A generic data management model.
Villazon – Terrazas et al. (2011)	(1) Specify; (2) Model; (3) Generate; (4) Publish; (5) Exploit	Create, Curate, Publish, Use	Focused on linked data	Not applicable in other contexts. Very generic.	Could be used from linked data publishers supporting re-use. Only for managerial purposes.
Hyland et al. (2011)	(1) identify, (2) model, (3) name, (4) describe, (5) convert, (6) publish	Pre-process, Curate, Publish	Focused on linked data publication process	Not applicable in other contexts. No inclusion of managerial processes and definition of a data plan.	Could be used from linked data publishers.
Hausenblas and Karnstedt (2010)	adding the steps (7) discovery (8) integration (9) use cases	Acquire, Process, Use	Focused on linked data. Includes re-use and the user side	Not applicable in other contexts	Could be used from linked data publishers and users.
Open Data Support Working Group	(1) Select (2) Model (3) Publish (4) Find (5) Integrate (6) Re-use (7) feedback	Create, Curate, Publish, Use, Feedback	Feedback loop Matching supply and demand	Very abstract. No peculiarities are addressed.	Could be used from linked data publishers and users. Could be used from public administrations for managerial purposes.

(continued)

Table 2.2 (continued)

Model	Key elements	Part of the open data life cycle covered	Strength(s) of this model	Weakness(es) of this model	Example of how this model can be used
van den Broek et al. (2011)	(1) identification, (2) preparation, (3) publication, (4) re-use and (5) evaluation	Pre-process, Curate, Publish, Use, Half Feedback step	The evaluation procedure	Not very descriptive	Could be used from linked data publishers supporting re-use and evaluation. Only for managerial purposes.
Auer et al. (2012)	Manual Revision and Authoring; Interlinking and Fusing; Classification and Enrichment; Quality Analysis; Evolution and Repair; Search and Browsing; Extraction; Storing and Querying	Create, Pre-process, Curate, Process, Use	Very detailed description of linked data manipulation	No feedback and collaboration mechanisms.	Could be used from public administrations providing linked data as well as linked data users.
Erl, Khattak, and Buhler (2016)	Data Identification; Data Acquisition and Filtering; Data Extraction; Data Validation and Cleansing; Data Aggregation and Representation; Data Modelling and Analysis; Data Visualization	Acquire, Curate, Process, Use	Very detailed description of big data handling from the user side	No publication procedures. More focused in the business sector and internal data analysis	Could be used from big data analysts and big data scientists
Kucera (2015)	OD Initiative initiation; Goal Setting; Publication Plan; Preparation of Datasets and infrastructure; Publication; Archiving; Evaluation.	Publication	Focused on managerial processes of data publication including evaluation procedures	Most for OGD initiatives	Could be used from public administration for publishing their data through an Open data initiative.

(continued)

Table 2.2 (continued)

Model	Key elements	Part of the open data life cycle covered	Strength(s) of this model	Weakness(es) of this model	Example of how this model can be used
Demchenko, Grosso, De Laat, and Membrey (2013)	Experiment planning; Data Collection and filtering; Data analysis (scientific data production); Data Re-purpose; Publication of data; Archive (data and scientific paper);	Acquire, Process, Use, Store	Actor blind/ Pro-cumers	Focused on Scientific Data Lifecycle	Could be used from universities embracing the open data paradigm for their research data and information.

https://joinup.ec.europa.eu/sites/default/files/D2.1.1%20Training%20Module%202.1%20The%20Linked%20Open%20Government%20Data%20Lifecycle_v0.11_EN.pdf

actor-interests are considered as "impediments" (Zuiderwijk, Janssen, Choenni, et al., 2012) or restrictions hindering an inherently good and beneficial idea (Meijer, de Hoog, Van Twist, van der Steen, & Scherpenisse, 2014). This is especially relevant as the different stakeholders involved have different understandings of and interests in open data which in turn influences the results (Janssen & Zuiderwijk, 2014; Zuiderwijk & Janssen, 2014a). Efforts have thus been made to develop more holistic analytic perspectives on open data e.g. based on complexity theory (Meijer et al., 2014) and the information ecology approach (Harrison, Guerrero, et al., 2012).

2.3 The Open Data Life Cycle: An Ecosystem Approach

The ecosystem perspective is widely used by scholars, policy makers and other stakeholders across different domains to discuss and explore the interdependencies among data, technology, actors and innovation in several organizational and technological contexts (Harrison, Guerrero, et al., 2012). The added value of the ecosystem perspective on open data is its focus on the relationships and interdependencies between the social (publishers and users of open data) and technological (data linking, big data analysis, storing, visualising) factors that affect the performance of open data activities within the life cycle (Dawes, Vidiasova, & Parkhimovich, 2016).

Addressing the new requirements under the ecosystem concept, a hybrid model has been produced incorporating steps from all its predecessors (see Sect. 2.2.4). Various steps addressing linked and big data specific capabilities along

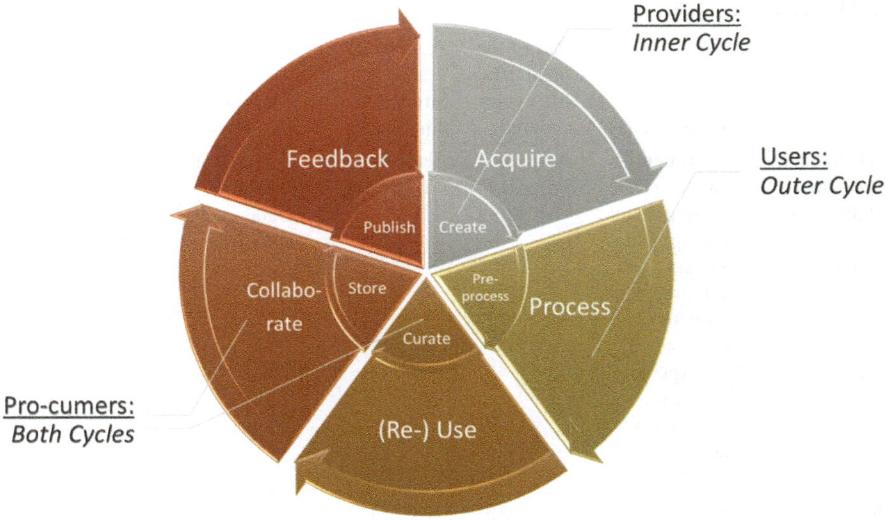

Fig. 2.1 The open data life cycle model

with the identification of the proper tools as well as the two different sides of the open data life cycle have been merged into a wider life cycle model providing the ecosystem view towards the achievement of the abovementioned impact from opening of public data. The curation life cycle is embedded in the "Curate" and "Pre-process" steps of the ENGAGE Open Data Life Cycle. Steps from the Open Data Publication Methodology (Kucera, 2015) have been also included. The basic development of the ENGAGE project since its conception is the collaboration step which is not included in any one of the above models. This is a result of the ENGAGE advanced functionality and web 2.0 capabilities which in fact provide a solid solution towards the realisation of the HORIZON 2020 vision concerning the e-infrastructures development for new workflows and collaboration.

Figure 2.1 introduces the Open Data Life Cycle Model. The different roles of the system are recognised in terms of inner and outer cycles. At this point we would like to clarify the pre-process step which is not referring to the calibration of data reducing their value. It incorporates the goal setting for each individual organisation publishing open data. The "Publish" step incorporates the publication planning which is related with the goals setting method of the "pre-processing" step. What is more, the feedback step refers to both the feedback from users as well as the assessment of the publication process against the goals setting.

Table 2.3 presents the methods and tools used for each life cycle stage regarding different types of data (big and linked).

Table 2.3 Methods and tools in each step of the open data life cycle

Life cycle stage	Tools	Methods
Create/Gather: The process of creating data	Sensors; RFID, IoT, IS; Human; Connection with already gathered open data; Hadoop for big data	Automated data creation (logs, network data) (Chen et al., 2014); Manual data entry; Linking with Open Data Portals
Pre-process: The managerial process of defining data quality	Detailed Metadata Standards; Evaluation Metrics and Models; Maturity Matrices; Unique identification (URIs and URLs)	Conceptualization & Goal setting; Evaluation plan and data quality; 3-layer Metadata Schema for portals
Curate: The process of meeting the required data quality and legal requirements	LOD Refine External Tool; Individual/Native Tools; R	Structuring; Anonymization; Metadata Refinement; Change Data Format; Data Cleansing
Store/Obtain: The decision making process of storing.	Data Centres; SPARQL Repositories for linked data; NoSQL & Document Databases for big data, linking with other datasets	Versioning; Data Linking; K-value and column oriented databases for big data (Chen et al., 2014)
Publish: The process covering legal issues	Upload Capability	Publication Plan Open Access Licensing Intellectual Property Rights
Retrieve/Acquire: The process of data acquisition through OD portals	OD portals (e.g. European data portal, world bank, national initiatives)	Multilingual search techniques APIs
Process: The process of data analysis	External data processing tools: Open Refine; R; Rapidminer; KNMINE; excel; Weka/ Pentaho	Data enrichment; Create Linked Open Data; Different Datasets combination; Text and Data Mining; Hashing; Cluster Analysis & Factor Analysis (Chen et al., 2014)
Use: The process of presenting the analysis outcomes	Internal & External Visualization tools; Statistical Packages; Linking with external artefacts (publications)	Statistical Analysis; Map Visualization; Chart Visualization; Plot Visualization; Visual Analytics; Cluster diagrams
Collaborate: The process of communicating with other data users	Collaboration space and workflow Web 2.0 capabilities and tools	Exchange notes/emails/ideas Create Groups of common interests
Feedback: The process of evaluating and providing feedback to data providers	Declare Need Web 2.0 Capabilities and Tools	Data Quality Rating; Requests on Open Data; Assessment of Publication

2.4 Different Uses of the Open Data Life Cycle

Much research has been conducted and many models have been designed in order to identify the open data life cycle as we can observe in Table 2.2. Each model focuses on different perspectives of open data regarding its nature (linked and big) and its purpose (data management, data curation). Even more research has been conducted for the definition of the data management life cycle (Committee on Earth Observation Satellites, Working Group on Information Systems and Services, 2011). This subsection analyses models that conceptualize the practices around handling data, from its generation to administrative practices involved in the provision of open data by public sector institutions to its use by third-parties.

This sub-section describes in more detail open data life cycle that best suits in different cases in order to illustrate specific aspects of the open data life cycle. As it could be discerned from the previous sub-sections the open data life cycle could be seen by two different perspectives. The major distinguishing aspect of the open data life cycle is the different stakeholders i.e. the publishers and the users. In the following sub-sections we present the open data life cycle from the publisher's side originating from the EU COSMODE project (Kucera, 2015) and the open data life cycle from the user's side. The user side consists of multiple stakeholders (i.e. scientists, journalists and citizens).

2.4.1 Towards Publication: The Data Publisher's Side

Open data are essential for achieving the United Nations' Sustainable Development Goals (The Open Working Group, 2015). Increased transparency, accountability and citizen participation (Jetzek, Avital, & Bjørn-Andersen, 2013), improved efficiency and effectiveness of public services (Huijboom, Broek, & Dutch Ministery of the Interior and Kingdom Relations, 2011), stimulation of economic growth; creation of social value (Gruen, Houghton, & Tooth, 2014) and positive impact on the quality and the effectiveness of the political debate (Ubaldi, 2013a), are only some examples of what our society could achieve through the opening and re-use of open data.

For the above-mentioned reasons, many countries all over the world design and implement OGD initiatives. Such initiatives have resulted in a greater availability of data including legislative interventions and development of digital infra-structures for this purpose (Commission of the European Communities, 2011). According to the Open Knowledge Network (2017), the "keep it simple" principle should be followed when opening up data. Even though OGD initiatives have been launched in many countries across the globe, only over 10% of the 1.290 datasets surveyed in the second edition of the Open Data Barometer study were published under an open license, in bulk and in machine-readable formats.

In addition, (Zuiderwijk, Janssen, Choenni, et al., 2012) observed that in practice it might be difficult to open up particular datasets because issues such as the confidentiality, data quality or the privacy infringement risks need to be addressed. Besides the privacy infringement risk, there might be other risks associated with the publication of OGD, such as publication of data against the law or possible misinterpretation of the data (Kucera & Chlapek, 2014). Ubaldi (2013a) points out that there are not only technical and legal challenges associated with the OGD initiatives but there are also challenges related to policy, financing, organization and culture. Chapter 4 provides a comprehensive overview of the organizational issues for opening up government data.

The abovementioned challenges and risks show that there is a need for an OGD publication methodology that would provide the responsible persons (publishers) with a clear guidance on how the OGD initiatives should be implemented and how the known challenges and risks should be addressed. If the challenges are not properly tackled it might prevent the expected benefits from being reaped (Ubaldi, 2013a). On the other hand, open data initiatives and practices take place in many different sectors, while users of open data often combine data from various domains.

In terms of the MePOD-VS methodology (Kucera, 2015) an Open Data initiative is an initiative executed by public sector bodies. Open Data publishing initiation might involve support of the top management of the public sector, and guarantee of departments and other stakeholders' participation. This is aligned with the SHARE-PSI 2.0 (2016) best practice on the *"Development of a Cross agency Strategy"*, which is presented in more detail in Chap. 4. According to (Moller, 2013), Open Data publication planning, Preparation of datasets and infrastructure, Open Data publication, cataloguing and maintenance and the Open Data archiving and retirement domains provide the necessary processes involved in the stages of the datasets lifecycle. Figure 2.2 illustrates the overall methodology and its process domains.

The main objective of the Open Data publication planning is to select a set of datasets for publication that is in line with the defined goals. The development of an open data publication plan will be used to steer the OGD initiative and it is aligned with the SHARE-PSI 2.0 (2016) best practice *"Open Data Publication Plan Development"*. Datasets planned to be released need to be prepared, e.g. they might need to be transformed into a suitable machine-readable format, enriched with metadata and properly licensed. Once the datasets are prepared they need to be made accessible and discoverable. Datasets and the respective metadata also need to be regularly updated (Lee, Cyganiak, & Decker, 2014).

Moreover, changes in legislation might affect what datasets particular public-sector organizations are able to publish as OGD, since the data could be characterized as private at some point after the beginning of the open data initiative. The Open Data archiving and retirement is part of the publication methodology in order to properly manage the end of the dataset lifecycle. Zuiderwijk et al. (2012a) have defined a process of selecting the data for publication. They argue that dealing with privacy-sensitive data, deletion policies, publishing after embargo periods instead of not publishing at all, adding related documents and adding information about the quality and completeness of datasets. The institutional context should be taken into account when using the guidance, as opening data requires considerable changes of organizations. Since the progress and impact evaluation of an OGD initiative is

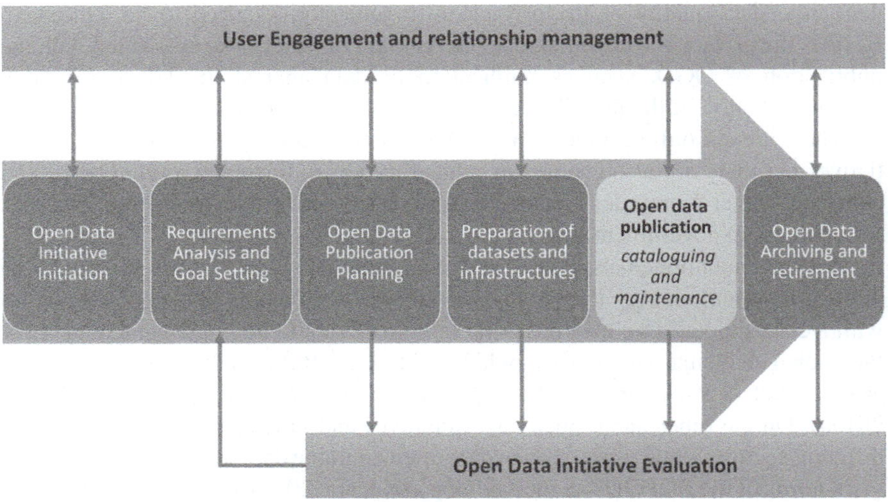

Fig. 2.2 Open data publication methodology, captured by Kucera (2015)

crucial for its development and implementation, a separate process domain is included dealing with the evaluation of progress against the Open Data publication plan and the defined goals.

User engagement and relationship management process domain is aimed at the identification of both actual and potential users of published data, the assessment of user's demands and requirements, as well as the setting up and execution of the communication strategy. It is also aiming at the assurance of feedback provision on the published data. While facilitation of the user feedback and re-use remains an important part of the OGD initiative this shift allows engaging users in the early stages of the OGD initiative which should help to establish a demand-driven release of data. This in turn should lead to a better alignment of data demand and supply.

Besides the tasks of the domains depicted in Fig. 2.2 there are other activities that need to be performed during the OGD publication such as the data quality management, benefits management or risk management (Nečaský et al., 2014). These topics are included as individual processes and not separate process domains. Since risk management and data quality should represent a continuous process, it is related to all process domains proposed in Fig. 2.2 in a way similar to the user engagement and relationship management process domain.

2.4.2 Towards Big Data Re-use: The Users' Side

Figure 2.3 presents a typical process of handling and processing big data in an enterprise environment beginning from the data identification towards data visualisation and utilisation of results.

Fig. 2.3 Big data user process adapted by Erl et al. (2016)

In a business environment the process starts with the identification of the problem to be tackled and the Key Performance Indicators (KPIs) that have to be measured determining the assessment criteria and guidance to the evaluation of analysis results. The problem to be solved should be quantified as a big data problem through the establishment of direct relations to one or more of the Big Data characteristics of volume, velocity, or variety. In Table 2.4 we describe the process step by step (Erl et al., 2016) and provide remarks on difficulties and crotchetiness for each one of them (Jagadish et al., 2014). Subsequent to analysis results being made available to business users to support business decision-making, such as via dashboards, there may be further opportunities to utilize the analysis results. After Data Visualization stage, it might be needed to determine how and where processed analysis data can be further leveraged. Depending on the nature of the analysis problems being addressed, it is possible for the analysis results to produce "models" that encapsulate new insights and understandings about the nature of the patterns and relationships that exist within the data that was analyzed.

2.4.3 Preparing a Scientific Data Infrastructure: Research Institutions

This subsection presents the user's perspective of the open data life cycle. As a user we have selected the researcher stakeholder. The constructors of the model begin with the statement that "Once the data is published, it is essential to allow other scientists to be able to validate and reproduce the data that they are interested in, and possibly contribute with new results" (Demchenko et al., 2013). Koop et al. (2011) argues that scientific data provenance should be taken into consideration by scientific data infrastructure providers.

Another aspect to take into consideration is to guarantee reusability of published data within the scientific community. Understanding semantic of the published data becomes an important issue to allow for reusability, and this had been traditionally being done manually. However, as we anticipate unprecedented scale of published data that will be generated in Big Data Science, attaching clear data semantic becomes a necessary condition for efficient reuse of published data. Learning from best practices in semantic web community on how to provide a reusable published data, will be one of consideration that will be addressed by the scientific data infrastructure. Big data are typically distributed both on the collection side and on the processing/access side: data need to be collected (sometimes in a time sensitive way

Table 2.4 Big data analysis process

Step	Description and remarks
Data Identification	Identifying a wider variety of data sources may increase the probability of finding hidden patterns and correlations. For example, to provide insight, it can be beneficial to identify as many types of related data sources as possible, especially when it is unclear exactly what to look for. Depending on the business scope of the analysis project and nature of the business problems being addressed, the required datasets and their sources can be internal and/or external to the enterprise. In the latter case, open data could be found from third-party data providers, such as data markets and publicly available datasets, are compiled. Some forms of open data may be embedded within blogs or other types of content-based web sites, in which case they may need to be harvested via automated tools.
Data Acquisition and Filtering	Depending on the type of data source, data may come as a collection of files, such as data purchased from a third-party data provider, or may require API integration, such as with Twitter. In many cases, especially concerning external, unstructured data, some or most of the acquired data may be irrelevant (noise) and can be discarded as part of the filtering process. Since the data filtered out for one analysis may possibly be valuable for a different type of analysis, it is advisable to store a copy of the original dataset before proceeding with the filtering. To improve the classification and querying, metadata (e.g. dataset size and structure, source information, date and time of creation or collection and language-specific information) can be added automatically from both internal and external data sources. It is vital that metadata be machine-readable and passed forward along subsequent analysis stages. This helps to maintain data provenance throughout the Big Data analytics lifecycle, which helps to establish and preserve data accuracy and quality.
Data Extraction	This step realizes the extraction of data from the sources according to the filtering criteria of the previous step. The required extent of extraction and transformation depends on the types of analytics and capabilities of the Big Data tool (i.e. extracting text for text analytics, which requires scans of whole documents, is simplified if the underlying Big Data tool can directly read the document in its native format).
Data Validation and Cleansing	Since invalid data can skew and falsify analysis results, this an important step of the process. Big Data can be unstructured without any indication of validity. Most data sources are notoriously unreliable: sensors can be faulty, humans may provide biased opinions, remote websites might be stale, and so on. Its complexity can further make it difficult to arrive at a set of suitable validation constraints. Understanding and modelling these sources of error is a first step toward developing data cleaning techniques. Provenance can play an important role in determining the accuracy and quality of questionable data.
Data Aggregation and Representation	This step deals with the required data reconciliation method to determine and represent the correct value. Data may be spread across multiple datasets, requiring that datasets be joined together via common fields, for example date or ID. In other cases, the same data fields may appear in multiple datasets, such as date of birth. The large volumes processed by Big Data tools can make data aggregation a time and effort-intensive operation. Future data analysis requirements need to be considered during this stage to help foster data reusability. A standardised data structure could act as a common denominator that may be used for a range of analysis techniques and projects. This can require establishing a central, standard analysis repository, such as a NoSQL database.

(continued)

Table 2.4 (continued)

Step	Description and remarks
Data Modelling and Analysis	The data analysis step is dedicated to carrying out the actual analysis task, which typically involves one or more types of analytics. This step can be iterative in nature, especially if the data analysis is exploratory, in which case analysis is repeated until the appropriate pattern or correlation is uncovered. Methods for querying and mining Big Data are fundamentally different from traditional statistical analysis on small samples. Big Data is often noisy, dynamic, heterogeneous, inter-related, and untrustworthy. Nevertheless, even noisy Big Data could be more valuable than tiny samples because general statistics obtained from frequent patterns and correlation analysis usually overpower individual fluctuations and often disclose more reliable hidden patterns and knowledge. In fact, with suitable statistical care, one can use approximate analyses to get good results without being overwhelmed by the volume.
Data Visualization	The last step of the process is to produce recognizable and useful insights through visuals to increase the value of the analysis of big data. The Data Visualization stage is dedicated to using data visualization techniques and tools to graphically communicate the analysis results for effective interpretation by business users. Users need to be able to understand the results in order to obtain value from the analysis and subsequently have the ability to provide feedback or make the right decisions. The results of completing the Data Visualization stage provide users with the ability to perform visual analysis, allowing for the discovery of answers to questions that users have not yet even formulated. The same results may be presented in a number of different ways, which can influence the interpretation of the results. Consequently, it is important to use the most suitable visualization technique by keeping the business domain in context. Another aspect to keep in mind is that providing a method of drilling down to comparatively simple statistics is crucial, in order for users to understand how the rolled up or aggregated results were generated.

or with other environmental attributes), distributed and/or replicated. Linking distributed data is one of the problems to be addressed by SDI. The required new approach to data management and handling in e-Science is reflected in the Scientific Data Lifecycle Management in Fig. 2.4, as a result of analysis of the existing practices in different scientific communities.

The generic scientific data lifecycle includes several consequent stages: research project or experiment planning; data collection; data processing; publishing research results; discussion, feedback; archiving (or discarding). The Scientific Data Lifecycle Management necessitates data storage and preservation at all stages what should allow data re-use and secondary research on the processed data and published results. However, this is possible only if the full data identification, cross-reference and linkage are implemented in scientific data infrastructure. Data integrity, access control and accountability must be supported during the whole data during lifecycle. Data curation is an important component of the scientific data lifecycle and must also be done in a secure and trustworthy way.

This example of scientific open data life cycle was selected based on its increased complexity compared to the two previous ones. The previous stakeholders do not

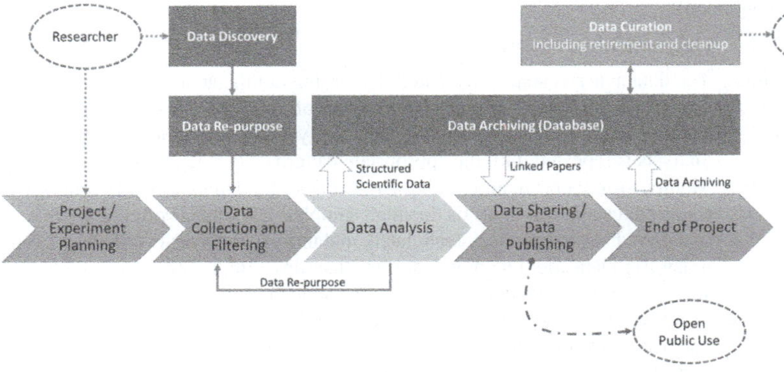

Fig. 2.4 Scientific data lifecycle management in e-science adapted from Demchenko et al. (2013)

pose so sophisticated requirements. Two are the most important issues regarding the peculiarities of this use case that are addressed by the open data life cycle model. Firstly, the recognition of the duality of a user to be both a user and a producer of data and secondly, the identification of the essential element of collaboration and interaction between different communities of users as well as between users and producers of data providing the necessary tools and workflows in the open data life cycle. These workflows will support the demand side of open data enhancing the exploitation step and closing the feedback loop.

2.4.4 Towards Linked Data Re-use: Publishers and Users

In order to support the full life cycle of linked open data, the Open Data Support Working Group resulted in the linked open data life cycle model presented in Fig. 2.5 including steps for both supply and demand (publishers and users) connecting them through the feedback step and thus closing the feedback loop.

In addition, the LOD2 stack is an integrated distribution of aligned tools which support the lifecycle of Linked (Open) Data from extraction to visualization and maintenance. The stack comprises tools from the LOD2 partners and third parties. With the ambition to identify these tools to support the creation and use of linked data, LOD2 project developed a more fine-grained 8-step life cycle model (Auer et al., 2012) formulated as follows: Extraction; Storing and Querying; Manual Revision and Authoring; Interlinking and Fusing; Classification and Enrichment; Quality Analysis; Evolution and Repair; Search and Browsing. Furthermore, LOD2 project has developed techniques for assessing quality based on characteristics such as provenance, context, coverage or structure. The open data life cycle presented in Sect. 2.3 has integrated these steps and tools incorporating the representation of linked data in the model, but this is not always the case. The LOD2 stack would guide better the manipulation of linked data since it is conceptualized and implemented targeting linked data specific characteristics. These specific characteristics towards data interoperability are mentioned and highlighted in Chap. 5.

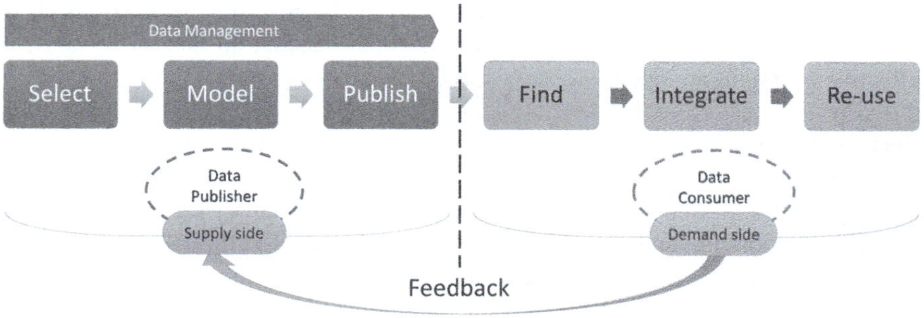

Fig. 2.5 OGD life cycle adapted from Open Data Support Working Group (https://joinup.ec.europa.eu/sites/default/files/D2.1.1%20Training%20Module%202.1%20The%20Linked%20Open%20Government%20Data%20Lifecycle_v0.11_EN.pdf)

2.5 Conclusions and Open Data Principles

This chapter identified the major data management and open data life-cycle models that exist in contemporary scientific literature. The major models have been presented in detail for each sub-category of technologies (linked data, big data) and associated stakeholders (publishers, users). Each life-cycle model could be used efficiently in different contexts. Finally, we introduced the new paradigm of the open data life cycle model from an ecosystem perspective including collaboration and feedback capabilities and acquainting with the notion of "data pro-sumer". A user with a possible dual role in the open data system being both producer and consumer of data.

The data itself is often treated as "a commodity rather than an artefact" (Meijer et al., 2014). However, how (open) data is understood and interpreted is shaped by the institutional and legal context, e.g. different perceptions of privacy and personal data. In a similar manner, some data can be considered more politicized than other. Also, different professional perspectives on data that refers to the same material object influence not only the sense-making, but the consideration of what data is actually important, the metrics of measurement etc. Altogether, this might even question the viability of a generic life-cycle model. Regarding the latter observation there should be an individual life-cycle model, which fits best in each situation.

Furthermore, this chapter identifies some principals for the open data that should be accompanying open data publication throughout its life-cycle. The principals for the open data publication process are:

Transparency-by-design (Janssen, 2015) Transparency-by-design refers to a principle where data about the functioning of government is automatically opened, can be easily accessed and interpreted, without being manipulated or being predefined or pre-processed. Transparency-by-design should ensure that information for effective public oversight is made available and that this information is clear and not ambiguous. Adherence to this principle requires that the mechanisms for creating transparency are integrated in the heart of the government functions. This does not necessarily imply that all data is opened, but that all data necessary for effective oversight are open.

Privacy-by-design (Janssen, 2015) Privacy-by-design means that systems and the governance of these systems, are developed to guarantee individual privacy. Privacy-by-design does not mean that data cannot be shared. Privacy-by-design should also contain measures to compromise privacy for the sake of national security. Peled (2014) argues that restrictions such as authorization from individuals before their medical data are released are required to increase data circulation. Although the need for privacy and transparency is intuitively clear, realizing both principles is a complex endeavour that might be one of the thorniest problems in digital government. Transparency and privacy are inter-dependent and non-dichotomous variables and complete transparency and privacy does not exist. Both principles compete with each other as well as with other principles underpinning our society and individual versus collective rights and responsibilities. Weighing transparency versus privacy requires a deep understanding of the situation at hand.

Quality-by-design The quality of data could be seen and assessed from different perspectives. The basic data quality measurements are: accuracy, completeness, consistency and timeliness. Even more perspectives could be included in the quality assessment, such as comprehensiveness, speed, security, correctness and others that will fully analysed in Chap. 8: Open Data Evaluation. Except the standard quality measures, data quality is heavily connected with the metadata provision, as well as the ascription of a persistent URI ensuring the unique identification of an open dataset. Furthermore, Tim Berners Lee introduces the 5-stars open data maturity model for quality measurement towards linked data focused mainly on the format of the provided data.

Closing the feedback loop One essential element of open data ecosystems concerns their development "through user adaptation, feedback loops and dynamic supplier and user interactions and other interacting factors" (Zuiderwijk et al., 2014). Open data ecosystems perform data production and usage-cycles with feedback loops, sharing of data back to publishers and also with the so-called infomediaries (Pollock, 2011). However, discussion and feedback loops appear barely to be part of existing open data practices and infrastructures. Zuiderwijk and Janssen (2013) found that after open data have been used, the provision of feedback to data providers or a discussion with them is quite important by not facilitated by existing open data infrastructures, though such mechanisms might be useful for improving open data quality, data release processes and policies. Dawes and Helbig (2010) found that such mechanisms can help users to obtain insight in how they can use and interpret open government data and generate value from them.

Besides generic policies and concepts on open data (Directive 2003/98/EC on the reuse of public sector information and the European Data Portal), various other – thematic – policies and concepts determine, guide or influence the provision, and the use of open data. In some domains the process towards openness is supported by legislative EU frameworks. In the geospatial / environmental data domain there are: (a) the INSPIRE framework Directive 2007/2/EC, (b) the Directive 2003/4 on public access to environmental information and (c) the earth observation with the EU Regulation 1159/2013 on the European Earth monitoring programme (GMES). In the transport domain there is the Directive 2010/40/EU on the deployment of Intelligent Transport Systems in the field of road transport. There is also a data model for statistical information (SDMX: the Statistical Data and Metadata eXchange) and a data model for social sciences study-level information (DDI - Data Documentation Initiative). In addition, in other domains – and across domains – initiatives have been taken and actions have been setup to support and enable open data. For some domains, this is strongly based on a national responsibility to promote transparency of government processes and products (e.g., access to legal data such as legislation, jurisprudence through national records acts). Particular effort has been made to promote and facilitate the opening of research and education data (e.g., European Commission 2016).

Best practices for open data have been defined and assigned to each element of PSI Directive on the re-use of open data from the SHARE-PSI 2.0[1] EU project and some more technical ones from the Data on the Web Best Practices Working Group (2017) of W3C. The next chapters will introduce the concept of open data analysed from technological business, socio-technical, operational, process, legal and governance perspectives, while the open data ecosystem will be largely described by its individual elements.

[1] https://www.w3.org/2013/share-psi/bp/

Chapter 3
Open Data Directives and Policies

> *"Currently a multiplicity of open data policies is under development at governmental agencies at various administrative levels, aiming to stimulate and guide the publication and use of data and to gain advantages from this."*

3.1 Introduction

In developing open data policies, organizations aim to stimulate and guide the publication and use of data and to gain advantages from this. Often open data policies are guided by a high-level directive, such as those of the United States (Obama, 2009b) and the European Commission (European Commission, 2013c). Open data policies are important, as their purpose is often to ensure the long-term availability of government information to create transparency and thereby to contribute to citizens' rights of public access to government information. This right is considered a fundamental tenet of democracy (Allen, 1992). Moreover, open data policies have the potential to increase the participation, interaction, self-empowerment and social inclusion of open data users (e.g. citizens) and providers alike, stimulating economic growth and innovation and realizing many other advantages.

Currently a multiplicity of open data policies is under development at governmental agencies at various administrative levels, such as policies being developed by the United Arab Emirates, Kenya, the region of New South Wales, the province of Utrecht in the Netherlands and the city of New York in the United States. Further developing the open data policy framework developed by Zuiderwijk and Janssen (2014a), this chapter explores the elements and characteristics of open data directives and policies. We look into the policy environment (context), the policy content (the policy input), policy implementation (performance indicators; the policy output), evaluation (public value realization; the policy impact) and policy change or termination (feedback). Furthermore, this chapter provides several examples of influential open data directives and policies that have been developed in the past two decades and it looks into the different levels (e.g. different administrative levels) at

Y. Charalabidis et al., *The World of Open Data*, Public Administration and Information Technology 28, https://doi.org/10.1007/978-3-319-90850-2_3

which open data policies have been defined. Subsequently, an in-depth case is provided concerning the development of the open data policy in The Netherlands. Finally, this chapter provides lessons learned from the development of open data policies that are useful for open data policy makers.

3.2 Policy: A Definition

A policy in general can be defined as "a purposive course of action followed by an actor or set of actors in dealing with a problem or matter of concern" (Anderson, 1990, p. 5). Policy deals with processes, activities and/or decisions that tackle societal problems (Stewart, Hedge, & Lester, 2008). Policies aim to achieve a certain impact in society and should include the factors that contribute to and influence this impact. Policies are developed using policy-making cycles which can consist of stages including problem identification and agenda setting, policy formulation, policy implementation and enforcement, policy evaluation, and policy change or termination (Stewart et al., 2008) (see Fig. 3.1).

In the first stage, the policy definition stage, the problem is identified and analysed. This results in the need to develop one or more policies. Moreover, the desired effects or outcomes, the scope, the target audience, and the timeline of the policy are often formulated. Subsequently, the actual policy is developed. This stage includes the problem analysis and the identification of alternative solutions. Thereafter, the implementation and enforcement stage start and the selected policy is implemented, for instance by adapting regulations or developing new services. Finally, the policy needs to be evaluated to ensure that the intended outcomes are realized. Based on the evaluation outcomes the policy may need to be changed or terminated. Then this whole

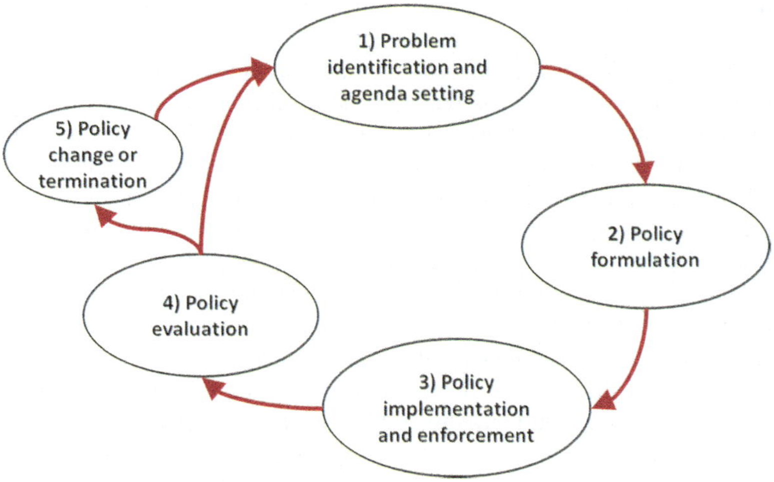

Fig. 3.1 Policy cycle. (Adapted from Stewart et al. (2008))

process starts all over again. Depictions of the policy process or policy stages vary through the literature and can be different per country and context. In addition, the order of the stages may differ. Policy development is often not a linear process and there are usually many iterations.

Policies, and particularly *open data* policies, are more than written documents in which intentions, choices and actions are described, as they define the broad open data regime of organizations and how they are realized and create their actual impact (Zuiderwijk & Janssen, 2014a). Following (Anderson, 1990, p. 5), we state that open data policies are *a purposive course of action followed by an actor or set of actors in dealing with open data-related issues*. This encompasses both dealing with issues related to the publication and related to the use of open data. Following Stewart et al. (2008), we state that *open data policy encompasses processes, activities and decisions that tackle open data related issues*. Open data policies can cover certain elements of the open data lifecycle or they can cover the complete lifecycle (see Chap. 2 about the open data lifecycle). When they cover the complete lifecycle, this means that they include the collection of data, the way that this data is opened and published, the place where it can be found, as well as how the data can be used and how feedback is dealt with. When they focus on a particular element, they can be focused on either obtaining access to data or on data protection or both. This is not always explicitly defined in a document but can also be an existing practice. For instance, we may consider the way that a governmental organization has been opening up its data in the past ten years a set policy, even if it is not explicitly described in a document.

3.3 Elements of Open Data Policies

Zuiderwijk and Janssen (2014a) developed a framework for comparing and evaluating open data policies (see Fig. 3.2). Based on the phases of the policy making cycle as defined by Stewart et al. (2008), they state that open data policies consist of the policy environment and context, the policy content (the input), performance indicators (the output) and public values (the impact). We extend this framework by adding open data policy change or termination as a fifth element.

The *contextual elements* of open data policies concern the open data policy environment. For example, this includes the regulatory context, the social context, and the political context. The contextual elements influence the *policy content*, including the policy strategy, the policy principles and practical aspects of opening data, such as the data quality and metadata provision. Policy content refers to the input for realizing societal values and contains the issues covered by the current open data policies. The combination of aspects that are part of the input of the open data process is expected to aim for a certain *output*. The policy output can be measured with performance indicators, such as the number of datasets opened up and the type of data use that takes place. Performance indicators can assist the *open data policy evaluation* and can show which *public value* is realized. Open data policies should

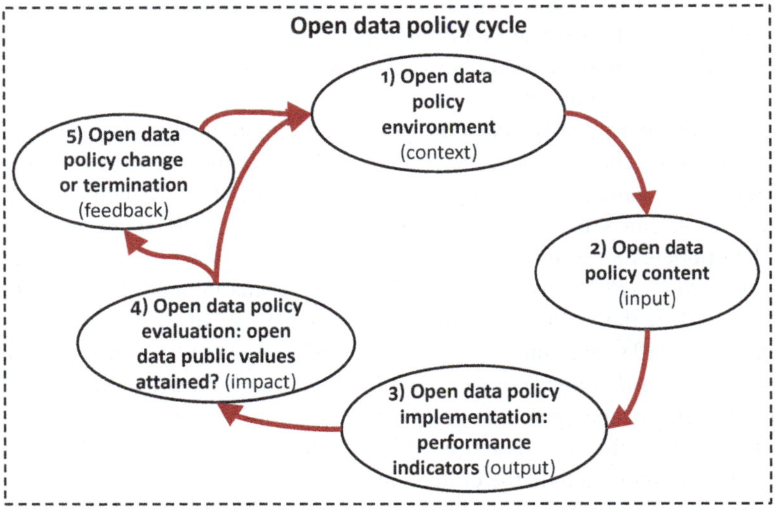

Fig. 3.2 Open data policy cycle. (Adapted from Stewart et al. (2008) and Zuiderwijk and Janssen (2014a))

not only focus on the opening of data, but they should pay special attention to improving the use of and value creation with open data. Policy evaluation should reveal the policy's impact on society, such as the creation of transparency and economic benefits. Finally, the evaluation will show whether the open data policy should be changed or terminated or not. *Feedback* on the policy may lead to policy improvements. Ideally, this cycle is iterated many times.

As policies are in a continuous state of flux, this framework can be viewed as a kind of policy-making cycle in which the created public values will influence the environment, context and policies. Below we will discuss each of the possible elements of open data policies using this framework. Note that open data policies are diverse and do not necessarily contain exactly these presented elements. Other elements and other orders are also possible.

3.3.1 Stage 1: Policy Environment (Context)

The first stage of the open data policy cycle concerns the policies' environment and its contextual aspects. In this stage, the problem is identified and agenda setting takes place, depending on the social, political, economic and regulatory context (see Fig. 3.3). The *social and demographic context* concerns the composition of the population, such as the age distribution, income, religion, behaviour, norms and values. The *political context* concerns the government structure, the government organization, and the way decisions are made. The *economic context* refers to the economic and financial situation, including the budget available for developing and

Fig. 3.3 Open data policy environment (context). (Adapted from Zuiderwijk and Janssen (2014a))

Stage 1: Policy environment (context)

- Social and demographic context
- Political context
- Economic context
- Legislation and regulatory context
- Culture and country
- Geographic level (e.g. country, region, city)
- Type of data providing organization(s)
- Mission type
- Key motivations and policy objectives
- Available resources (ICTs, human resources)
- Available open data platform
- Resource allocation

implementing the open data policy. The *legislation and regulatory context* comprises the laws and regulations that need to be taken into account when developing the open data policy, such as European open data directives and the Open Government Law in the Netherlands ('Wet Open Overheid' in Dutch). Developers of open data policies need to take into account the legislation that the policy is related to, and they may refer to this in an open data policy document.

Problem identification and agenda setting are also influenced by other contextual aspects, such as the *existing (organizational) culture* (e.g. the level of individualism and collectivism, power distance, and long term/short term orientation (see Hofstede, 2001)) and the *geographical level* (e.g. the country or city in which the policy is developed or the objectives of the organization that develops the policy). Furthermore, open data policies often include the *type of data providing organization(s)*. Some open data policies are created for a large range of organizations (e.g. a country's national open data policy), whereas other open data policies are specific to a particular organization (e.g. a ministry).

In the *mission* of these organizations open data can be, for instance, regulatory, strategic, or a social service.

- *Regulatory.* Opening data regulatorily may concern an organization that opens up data because it is forced to do so according to national or international legislation. For instance, a museum or library may be forced to open up (part of) its data because of the European PSI-directive or a national open data policy.
- *Strategic.* Opening data strategically concerns opening up data for the purpose of showing how transparent the organization is, to enhance trust of citizens or clients, or for obtaining feedback on the data collected by an organization to subsequently improve the quality of the data or the quality of work processes. For instance, as an example outside of the government context, Nike opens up factory, footprints and materials data that gives insights in the working processes of the company. This should enhance monitoring effectiveness and improve workers' conditions (Houk, 2011).
- *Social service.* Data provision as a social service may concern an organization that aims to open up data to create a more effective organization, build a stronger community or promote new opportunities. For example, a national government

may open up its data to build a community of entrepreneurs that have equal access to open data and that can use open data to develop new business models.

Open data policies may contain these types of missions, as well as the *key motivations and policy objectives* for opening data. The motivations and objectives can be on a high level of abstraction, such as innovation, transparency, participation of citizens, and economic value creation, or they can be more specific, such as providing a certain type of data to a certain community so that useful applications can be developed for a certain target group.

Other contextual factors influencing the development and design of open data policies include the available Information and Communication Technologies (ICTs), such as an appropriate internet infrastructure, *open data platforms* and Application Programming Interfaces (APIs), but also the *availability and allocation of resources* such as skilled personal for making data available and providing data is a useful format. Open data policies sometimes define the resources that are needed for opening and using data, or even the budget that is available for this. The open data policy may also give information regarding where the data is published, for instance, on a national open data portal.

3.3.2 Stage 2: Policy Content (Input)

In the second stage of the open data policy cycle the content of the open data policy is defined. This stage consists of a number of key elements, some of which are more related to the data opening processes and others which are more related to data management (see Fig. 3.4).

Stage 2: Policy content (input)

Data opening processes
- Policy strategy and principles for opening data
- Actors involved in opening data
- Targeted open data users
- Types of data opened and not opened
- Policy measures and instruments
- Provision of (technical) support for opening data
- Provision of (technical) support for open data use
- Type of engagement of and interaction between data providers and users
- Promotion of data and data use

Data management
- Data processing before opening
- Data quality
- Selected open data license and use conditions
- Numbers or percentages of opened datasets
- Data and metadata provision
- Data access and availability (e.g. required registration, portal)
- Way of presenting data and metadata to users (e.g. formats, standards)
- Data and metadata update frequency

Fig. 3.4 Open data policy content (input). (Adapted from Zuiderwijk and Janssen (2014a))

3.3.2.1 Data Opening Processes

The open data policy content concerning data opening processes includes the *policy strategy and principles for opening data*. This strategy and these principles sketch the outlines of the way the policy is intended to work after implementation. For instance, data may be opened only to certain target groups, or to any user. Another principle is that data is open by default, which means that the data is opened by default, unless there are significant barriers such as privacy aspects or data sensitivity. Open data policies may also include the *actors involved in opening data*, such as the parties involved in opening up data and the parties involved in publishing the data on open data platforms. Open data policies may describe the *typical open data users that are targeted*. This can be done in a detailed level (e.g. technically-skilled application developers in the areas of geographic information or academic researchers in the social sciences domain) or on a high level (e.g. citizens, developers or researchers).

Open data policies may contain the *types of data that are not opened*, such as incomplete data, data that is sensitive to misuse, and policy-confidential data, and they may make explicit or give examples of the *types of data that is opened*, such as data on certain topics or from certain registers. Open data policies describe the *measures and instruments* that are used to develop and evaluate the policy, such as websites, letters, speeches, networks, and social media. Other examples of such measures and instruments are fines and rewards, that can be used to stimulate data opening, for example by having a policy that requires departments within the organization to explain if a certain condition of the policy cannot be met. Open data policies can also describe multilateral instruments, such as contracts, to stimulate data opening.

Some open data policies provide information concerning the *technical and non-technical support that should be given to data providers and to data users*. For instance, data providers may be supported by a data steward who can explain or check whether data protection legislation would be violated if a certain dataset would be opened. Data users may be supported via support tools on the open data portal, via e-mail, and via social media. Open data policies may discuss the *type of engagement* that is envisioned between the data provider and the data user. There may be much interaction and feedback processes could be institutionalized, this may be lacking completely or there may be some level of engagement and interaction in between. The open data policy defines *whether data use is promoted* to potential new open data users and how this is done. For instance, data use can be encouraged through the organization and advertisement of hackathons and app contests.

3.3.2.2 Data Management

The open data policy content concerning data management includes the type and amount of *data processing required before opening the data*. Data is often stripped of personal details and checked in terms of *quality*, including its validity, anonymity, reliability, completeness, representativeness and documentation, before it is opened. The way in which data is processed often influences under which conditions the end-user can use the data and which *licenses and use conditions* may be needed. For example, if a dataset is completely anonymized and aggregated and the data collection process is well-documented, the user may receive more freedom in reusing the data then for a dataset that contains "rawer" (i.e. unprocessed) data. Open data policies need to define which licenses will apply to the use of the data, as well as the type of information that the user needs to provide before downloading the data. Examples of open data licenses are e.g. the Open Government License UK, Creative Commons (Petychakis, Vasileiou, Georgis, Mouzakitis, & Psarras, 2014) and Open Data Commons (Miller, Styles, & Heath, 2008).

Furthermore, the open data policy encompasses the *number, types or percentages of opened and non-opened datasets* and their related *metadata*, although numbers and types do not say anything about the usefulness and quality of the data. Although this is difficult to measure, the policy can contain a statement about the quality that the data should have when it is collected and before it is opened. Open data policies include *the way that the access* to the data is given. For instance, they show whether the user needs to register or whether the users should accept certain use conditions before the dataset can be downloaded. It also concerns the *data availability*, including the portal where the data can be found. Moreover, the policy content defines the *way of presenting data and metadata to users*, including the technical standards and formats for open data (e.g. CSV or XLS). It refers to the type of metadata that is provided with the data, such as descriptive, contextual and detailed metadata (Jeffery, Asserson, Houssos, & Jörg, 2013; Zuiderwijk, 2015a), as well as the standard that is used to provide the metadata (e.g. CERIF, CKAN or DC) (see Chap. 5). Finally, open data policies include the *frequency of updating data and metadata*.

3.3.3 Stage 3: Policy Implementation: Performance Indicators (Output)

In the third phase of the open data policy cycle, the policy is implemented and enforced. The performance indicators of the open data policy are defined. Performance indicators can be used to evaluate the progress of an open data policy at the fourth stage of the policy making cycle. The policy ideally contains metrics, such as indicators for output steering. Based on the developed policy objectives, indicators may be developed concerning the provision of the data, the use of the data or a combination of those (Susha, Zuiderwijk, Janssen, & Grönlund, 2015) (see Fig. 3.5).

Performance indicators concerning the provision of open data focus primarily on which data is available and in which form. As an example, the Open Data Index produced by the Open Knowledge Foundation focuses on concepts related to data provision, namely: publicly available data, freely available data, data available online, data in machine-readable formats, data available in bulk, up-to-date data, open license, available terms of use, metadata and data quality. Another example concerns the set of open data guidelines created by the Sunlight Foundation. It addresses what data should be public, how to make data public, and how to implement the open data policy (Sunlight Foundation, 2014). This includes principles con-

Stage 3: policy implementation: performance indicators (output)

Performance indicators concerning open data provision, for example:
- The number of datasets opened
- The machine-readability of data
- The metadata provided for each dataset

Performance indicators concerning open data use, for example:
- The number of data users
- The number of dataset downloads
- Type of data use

Fig. 3.5 Open data performance indicators (output). (Adapted from Zuiderwijk and Janssen (2014a) and Susha et al. (2015)

cerning machine-readable formats, the creation of data portals that should provide easy access, and the requirement of publishing metadata (see Chap. 5). The open data policy may include performance indicators concerning data provision such as those provided by the Open Data Index and the Sunlight Foundation.

Performance indicators should not only be focused on the provision of the data, as its use is also of critical importance. *Performance indicators for open data use* focus on actual data use and users. Performance indicators in this area consider numbers and characteristics of open data users, the way that the opened data is used and feedback and interaction between open data users and providers. Since open data is made available to *any* user, the data provider often does not have insight in who uses the data, which complicates setting performance indicators for data use and evaluating to which degree those indicators have been met. Open data use performance indicators usually give a limited view of actual data use. For instance, data users may not be interested in providing feedback concerning the way in which they used a dataset to the data provider, and the number of dataset downloads does not reflect the way in which open datasets have been used.

3.3.4 Stage 4: Evaluation: Public Value Realised? (Impact)

Data providers often want to know the successfulness of their implemented open data policy, which requires evaluation. Ultimately, open data policies meet the set performance indicators. Beyond performance indicators, they realize the benefits that they aim for, contribute to public values and have a large impact on society. The evaluation of impact can be assessed per open data policy, yet it is difficult to assess whether a certain impact has been caused by a certain open data policy. Impact assessment is therefore often focused on consolidating impact evidence from multiple open data policies on a larger scale. The evaluation of implemented open data policy is further complicated as many different stakeholders are involved (e.g. policy makers, data providing organizations, data users) and success may have a different meaning to them.

Stage 4: public value (impact)

Political and social value, for example:
- Increased transparency
- Increased participation
- Increased democratic accountability
- Increased trust in government

Economic value, for example:
- Stimulated innovation
- Economic growth
- Greater efficiency of government

Technical and operational value, for example:
- Ability to reuse data
- Fair decision-making by enabling comparison of different sources
- Easier discovery of data

Fig. 3.6 Open data public value creation (impact). (Adapted from Janssen, Charalabidis, and Zuiderwijk (2012))

Evaluation of realized public value can be done against the objectives set at the first stage of the policy cycle or data providing organizations may be compared to one another through benchmarking. Figure 3.6 provides several examples of open data policy impact. This impact can be in different areas, such as political, social, economic, operational and technical (Janssen et al., 2012).

- *Political and social value.* For instance, open data policies aim to create political and social value by increasing transparency (Kulk & van Loenen, 2012; Welle Donker, van Loenen, & Bregt, 2016; Zuiderwijk, 2015a), increasing participation (Evans & Campos, 2013; Lathrop & Ruma, 2010), increasing democratic accountability (Harrison, Guerrero, et al., 2012), stimulating knowledge development (Chun, Shulman, Sandoval, & Hovy, 2010) and increasing trust in government (Linders, 2013).
- *Economic value.* Examples of economic value include stimulated innovation (Lee & Kwak, 2012; Ubaldi, 2013b), economic growth (Arzberger et al., 2004; Bertot, Jaeger, & Grimes, 2010), greater efficiency of government (Kassen, 2013; Moon, 2002; Welle Donker et al., 2016), and access to external problem-solving capacity and resources (Harrison, Pardo, & Cook, 2012).
- *Technical and operational value.* Examples of operational and technical value concern the ability to reuse data (Ubaldi, 2013b; Yu & Robinson, 2012), fair decision-making by enabling comparison of different sources (Harrison, Guerrero, et al., 2012), easier discovery of data (Villazón-Terrazas, Vilches-Blázquez, Corcho, & Gómez-Pérez, 2011), contribution towards the improvement of administrative processes (Coglianese, 2009; Harrison, Guerrero, et al., 2012; Welle Donker et al., 2016) and use of the wisdom of the crowds: tapping into the intelligence of the collective (Lathrop & Ruma, 2010).

Several benchmarks to evaluate open data policy impact have been developed so far. An example of the evaluation of open data policy impact is the Open Data

Barometer survey carried out by the Web Foundation (Davies, 2013). It uses a crowd sourced survey to assess political, economic and social impacts. Other examples of evaluating impact include analysing log data to obtain more insight in who uses open data (Van Loenen, Ubacht, Labots, & Zuiderwijk, 2017) and creating a network of data providers and companies using open data by the Open Data 500 project, showing which companies use open government data from which sector and from which governmental organization in the United States (GovLab, 2014).

Each benchmark has a different scope, different strengths and weaknesses, and can be used to evaluate different elements of open data policies (Susha et al., 2015). The benchmarks can complement each other (idem). Many benchmarks focus on national open data policies, whereas local, regional and international policies are also under development and need to be evaluated.

3.3.5 Stage 5: Policy Change or Termination (Feedback)

The evaluation of open data policies (e.g. through benchmarks) should provide support for improving the existing situation (Susha et al., 2015). Based on the outcomes of the previous stages in the policy making cycle, open data policies can be changed or even terminated. As the field of open data is progressing rapidly, it is important to continuously evaluate the value generated through open data policies and to identify areas for improvement (Susha et al., 2015).

3.4 Directives Promoting Open Data Policy Development

In this section we provide an overview of directives that promote the development of open data policies. As explained before, in addition to published documents, open data policies also concern existing practices. Stimulated by various directives, many open data policies have been developed worldwide. For example, a report of the European Data Portal (2016c) shows that 25 out of the 31 European countries that they investigated have developed a national open data policy. Open data policies that are often seen as important for the boost of the open data movement include those of the European Commission, the United States of America and the Open Government Partnership.

3.4.1 European Commission DIRECTIVE 2003/98/EC

DIRECTIVE 2003/98/EC by the European Commission, the so-called Public Sector Information (PSI) directive, is often seen as the starting point (European Commission, 2003). This document provides "a general framework to ensure fair, proportionate and non-discriminatory conditions for the re-use of PSI". It states that "Member States shall ensure that, where the re-use of documents held by public sector bodies is allowed, these documents shall be re-usable for commercial or non-commercial purposes" (idem, p. 5). For most European countries their open data policy is similar to the Public Sector Information policy, which is mostly based on the transposition of the revised European PSI Directive. The Directive covers not only written texts, but also databases, audio files and film fragments. It excludes educational, scientific, and broadcasting sectors (European Commission, 2017).

DIRECTIVE 2003/98/EC by the European Commission was complemented by directives and policies in specific sectors (European Commission, 2011c), such as those concerning:

- access to open environmental data (European Commission, 2007, 2016);
- access to open marine data (European Commission, 2010b);
- access to data concerning innovative transport technologies (European Commission, 2010c); and
- access to data concerning cultural heritage material and digital libraries (European Commission, 2011a).

These directives are developing over time and are updated regularly. They provide a general framework to member states for making available particular types of data. For instance, DIRECTIVE 2007/2/EC establishing an Infrastructure for Spatial Information in the European Community (for short, the INSPIRE directive) directs the creation of an infrastructure for spatial information. The above-mentioned directives are often generic without specifying how the envisioned results should be achieved. They provide guidelines or a high-level framework for the development of (more specific) policies.

In 2011, the European Commission updated its open data strategy (European Commission, 2011c). Compared to the 2003 Directive on the re-use of public sector information the following changes were made:

- It was made "a general rule that all documents made accessible by public sector bodies can be re-used for any purpose, commercial or non-commercial, unless protected by third party copyright" (European Commission, 2011c, p. 1);
- The principle was established that "public bodies should not be allowed to charge more than costs triggered by the individual request for data (marginal costs)" (European Commission, 2011c, p. 1) meaning that most data should be offered for free;
- It was made "compulsory to provide data in commonly-used, machine-readable formats, to ensure data can be effectively re-used" (European Commission, 2011e, p. 1);
- These principles were enforced by ensuring regulatory oversight, and also libraries, museums and archives were then included in the reach of the directive (European Commission, 2011e).

Moreover, the European Commission promised to publish its own data through a portal that serves as a single-access point for open data from all EU institutions, bodies and agencies and national authorities. Former European Commission Vice President Neelie Kroes endorsed this open data policy. She stated: "*We are sending a strong signal to administrations today. Your data is worth more if you give it away. So start releasing it now*" (European Commission, 2011e). The European Parliament formally adopted the amended EU open data policy in June 2013 (European Commission, 2013a).

3.4.2 U.S.A. Memoranda and Directives

In 2009, U.S. President Obama signed and published a Memorandum on Transparency and Open Government (Obama, 2009a). The memorandum is targeted at the American heads of executive departments and agencies. Obama stated: "my Administration is committed to creating an unprecedented level of openness in Government" and "the government should be transparent, participatory and collaborative" (Obama, 2009a, p. 1). In this memorandum the president instructed the Director of the Office of Management and Budget to issue an Open Government Directive, which became available later that year. The Open Government Directive directs executive departments and agencies to take specific actions to implement the principles of transparency, participation, and collaboration as described in the President's Memorandum, and sets specific deadlines (Obama, 2009b). The directive states that executive departments and agencies should take four main steps toward creating a more open government:

- publish government information online;
- improve the quality of government information;
- create and institutionalize a culture of open government; and
- create an enabling policy framework for open government.

Agencies should produce an action plan that specifies which actions they will undertake to achieve this and by when they will do so.

In 2012, the Obama administration developed a Digital Government Strategy – Building a twenty-first century Platform to Better Serve the American People (Obama, 2012a). This strategy is characterized by Obama's statement: "I want us to ask ourselves every day, how are we using technology to make a real difference in people's lives." The President states that a digital American government should be efficient, effective and focused on improving the delivery of services to the American people. This should be realized by enabling "citizens and an increasingly mobile federal workforce to securely access high quality digital government information, data and services – anywhere, anytime, on any device" (idem, p. 27). Furthermore, to assure that the government adapts to this new digital world, a modern infrastructure should be provided to support digital government efforts and to reduce costs (Obama, 2012a).

3.4.3 *Other Directives and Guidelines for Open Data Policy Development*

Several other important international initiatives that promote open data policy development include the following.

3.4.3.1 Open Government Partnership (OGP)

The Open Government Partnership (OGP) was launched in September 2011 by governments from eight countries (Brazil, Indonesia, Mexico, Norway, the Philippines, South Africa, the United Kingdom and the United States). These countries endorsed the Open Government Declaration and announced their action plans to make their governments more open. In addition to these 8 countries, 67 national governments and 15 subnational governments have joined the OGP since its launch in 2011. Each of them develops a country action plan through public consultation and endorsed the high-level Open Government Declaration. OGP aims at defining concrete government commitments to stimulate transparency, empower citizens, fight corruption, and harness new technologies to strengthen governance (Open Government Partnership, 2017).

3.4.3.2 Open Data Charter

In 2013, the G8 leaders signed an Open Data Charter, consisting of five main principles. All nations involved agreed to establish an expectation that government data should be published openly by default (European Commission, 2013e). Various groups from governments, multilateral organizations, civil society and private sector (including the OGP Open Data Working Group) collaborated to develop the principles further in the following years (Open Data Charter, 2017). In 2015, they agreed on an international Open Data Charter, with six principles for the release of data:

1. Open by Default;
2. Timely and Comprehensive;
3. Accessible and Useable;
4. Comparable and Interoperable;
5. For Improved Governance and Citizen Engagement; and
6. For Inclusive Development and Innovation.

These principles ultimately support open data use. The International Open Data Charter has already been adopted by 47 governments (17 national and 30 local/subnational – as of August 2017). The Charter recommends standardisation of data and metadata, stimulates cultural change, promotes engagement with citizens and civil society and encourages increased attention for data literacy, training programs and entrepreneurship (Open Data Charter, 2017).

3.5 Examples of Open Data Policies at Different Levels

Currently a multiplicity of open data policies and directives are under development at governmental agencies at various administrative levels. Table 3.1 depicts some examples of developed open data policies and directives at international, national, state, regional and local/city level. The final column, containing references to the policy/directive, is also an example. Usually a policy is not described in one single document, but information about the actual policy needs to be obtained from multiple sources. The policies are diverse and support open data publication and use in different ways. From the table below, we can conclude that open data policies are under development all over the world and at a variety of administrative levels.

Table 3.1 Examples of developed open data policies and directives

Open data policy/directive level	Geographical area that the open data policy/directive applies to	Example of developed open data policy/directive	Reference to policy/directive
International	European Commission	DIRECTIVE 2003/98/EC	European Commission (2003), European Commission (2013c)
	United Arab Emirates	Open Data Policy	United Arab Emirates – Federal Customs Authority (2016)
National	India	Open Data Policy (NDSAP) of India	Digital India (n.d.)
	Brazil	Practical manual of the Transparency Portal of the Federal Government	Governo Federal (2010)
	Kenya	Government of Kenya open data initiative	Kenya ICT Board (2017)
State	New South Wales, Australia	Open data policy	State of New South Wales – Department of Finance (2016)
Regional	Province of Utrecht, the Netherlands	Utrecht Open Data	Province Utrecht (2017)
	Catalonia, Spain	Partnership Agreement between the Government of Catalonia and the Wikimedia Amical association	Generalitat de Catalunya (2017)
Local/city	New York, U.S.A.	Open data policy and technical standards manual	City of New York (2016)
	Chicago, U.S.A.	Open Data Executive Order (No. 2012–2)	City of Chicago (2012)

3.6 Use Case: The Dutch Open Data Policy

In this section we used the elements of open data policies as described at the beginning of this chapter to analyse the national open data policy of the Netherlands. This policy has been described in a variety of documents, complemented with information obtained from open data portals, discussions with civil servants responsible for Dutch open data policies at different levels and organizations, and practical experience. Table 3.2 depicts the main characteristics of the Dutch national open data policy.

The social, political, economic, and regulatory context shape the Dutch open data policy. Policymaking in the Netherlands is consensus-based (Pollitt & Bouckaert, 2011). Politt and Bouckaert write that, compared to other countries, "Dutch ministries are relatively open organizations" (p. 271). This is influenced by the Dutch system that allows for consultative and advisory councils (Pollitt & Bouckaert, 2011). The Netherlands is a decentralized unitary constitutional state based on a parliamentary democracy (Pollitt & Bouckaert, 2011). The Netherlands has a Gross Domestic Product (GDP) of 770.845 billion dollar in 2016, compared to for instance 18.596 trillion in the United States and 2.619 trillion in the United Kingdom (The World Bank, 2016).

Several strategies, laws, letters, action plans and vision statements form the regulatory context of the Dutch open data policy. The EU strategy forces the development and implementation of a national open data policy (European Commission, 2013c). In addition, a National Open Data Agenda has been developed (Ministerie van Binnenlandse Zaken en Koninkrijksrelaties, 2016). Legislation that has been developed in this area includes:

- Law Reuse of Government Information – Wet Openbaarheid van Bestuur. Opening data on request, Freedom of Information Legislation.
- Law Openness of Public Administration – Wet Hergebruik Overheidsinformatie). Actively opening data.
- Law Open Government (Wet Open Overheid) – currently handled by the Upper House of Dutch Parliament.

The Netherlands has joined the Open Government Partnership and developed an action plan (Ministerie van Binnenlandse Zaken en Koninkrijksrelaties, 2013a), a Vision Open Government (Ministerie van Binnenlandse Zaken en Koninkrijksrelaties, 2013b) and the Minister of the Interior sent the Second Chamber several letters concerning the government's open data policy (Ministerie van Binnenlandse Zaken en Koninkrijksrelaties, 2017c). All these documents contain information concerning the elements of the Dutch national open data policy.

Furthermore, the policy environment of the Dutch open data policy is characterized by a population of about ~17 million inhabitants. Cultural characteristics concern the low power distance (being independent, hierarchy for convenience only, equal rights, direct and participative communication), a relatively individualist society (loosely-knit social framework of individuals), a relatively feminine society

Table 3.2 Policy environment characteristics of the Dutch open data policy

Policy elements		National open data policy of the Netherlands
Stage 1: Policy environment	Social context	Policymaking is consensus-based and governmental organizations are relatively open (Pollitt & Bouckaert, 2011)
	Political context	Decentralized unitary constitutional state, based on a parliamentary democracy (Pollitt & Bouckaert, 2011)
	Economic context	GDP: 770,845 billion dollar in 2016 (The World Bank, 2016)
	Legislation and regulatory context	EU strategy (European Commission, 2013c) National Open Data Agenda (Ministerie van Binnenlandse Zaken en Koninkrijksrelaties, 2016) Laws (including the Law Reuse of Government Information, Law Openness of Public Administration and Law Open Government (the latter is under review) Open Government Partnership (OGP) Action plan for OGP (Ministerie van Binnenlandse Zaken en Koninkrijksrelaties, 2013a) Vision Open Government (Ministerie van Binnenlandse Zaken en Koninkrijksrelaties, 2013b) Letters sent by the Minister of the Interior to the Second Chamber (Ministerie van Binnenlandse Zaken en Koninkrijksrelaties, 2017c)
	Culture and country	~17 million inhabitants. Cultural characteristics: low power distance, individualist society, feminine society, slight preference for avoiding uncertainty (Hofstede, 2001; Hofstede, Hofstede, & Minkov, 2010; Hofstede Insights, 2017)
	Geographic level	Country (national)
	Type of data providing organizations	Ministries, provinces, municipalities, and other governmental organizations
	Key motivations and policy objectives	Open data is beneficial to the society Open government data stimulate private organizations, innovation, new business models and employment Insights in the available data and information of the government can contribute to cost reductions and improving policy processes (Ministerie van Binnenlandse Zaken en Koninkrijksrelaties, 2016, p. 1)
	Mission type	Mainly strategic, focus on transparency and democratic accountability (Ministerie van Binnenlandse Zaken en Koninkrijksrelaties, 2016)
	Available resources	Human resources and IT resources (Ministerie van Binnenlandse Zaken en Koninkrijksrelaties, 2016)
	Available open data platform	One national open data portal has been developed: data. overheid.nl At the same time various other open data portals are available, e.g. for specific ministries or domains (e.g. geographical data or social science data).
	Resource allocation	Human resources: at the national level to support the opening process (for questions concerning technology, organization and licenses) IT resources: a national portal

(important to keep the life/work balance) and a slight preference for avoiding uncertainty (Hofstede, 2001; Hofstede et al., 2010; Hofstede Insights, 2017).

The national open data policy is developed at the central level of government, under responsibility of the Ministry of the Interior and Kingdom Relations, yet other governmental organizations, including ministries, provinces and municipalities are also developing their own policies. At the national level, the policy is mainly strategy, as it focuses on transparency and democratic accountability (Ministerie van Binnenlandse Zaken en Koninkrijksrelaties, 2016). Key motivations and policy objectives are: "The society can profit from open data. Governmental data stimulate private organizations and stimulate innovation, new business models and employment. Insights in the available data and information of the government can contribute to cost reductions and improving policy processes." (Ministerie van Binnenlandse Zaken en Koninkrijksrelaties, 2016, p. 1).

Human resources are available at the national level to support the opening process (for questions concerning technology, organization and licenses). Regarding available Information Technology (IT) resources, a national portal is available, namely data.overheid.nl. Yet, many organizations and domains develop their own portals (e.g. one portal for geographical data and one portal per municipality), and various datasets are available at multiple places. For instance, open data portals are available for specific ministries and domains (e.g. geographical data or social science data) (Table 3.3).

The policy content is first characterized by the policy strategy and principles. The basic principle of the Dutch open data policy is to open data by default (Ministerie van Binnenlandse Zaken en Koninkrijksrelaties, 2016). Each department is responsible and accountable for the execution and approach of opening its data, coordinated under the supervision of the ministry of the Interior and Kingdom Relations (idem). The main actors involved in developing the Dutch open data policy are governmental organizations collecting and creating data and Information Technology (IT) providers. Targeted users are particularly citizens and entrepreneurs, although anyone can use government data. Through the national portal (data.overheid.nl) data is made available to users concerning a variety of themes: administration, culture and recreation, economy, finance, housing, international, agriculture, migration and integration, nature and environment, education and science, public order and safety, law, space and infrastructure, social security, traffic, work, care and health. Privacy sensitive data, other sensitive data and other data that is not appropriate for opening remains closed. Regarding the open data measures and instruments, the Dutch national open data policy defines three focus areas:

- Incentivisation and disclosure of datasets – focused on numbers and prioritization of datasets
- Progress monitoring and quality. Contains measures to monitor the quality of the metadata and the progress of disclosing data.
- Supporting the disclosure, technology and users – offers help to data managers. Collects wishes and questions of data users (Ministerie van Binnenlandse Zaken en Koninkrijksrelaties, 2016).

Table 3.3 Policy content characteristics of the Dutch open data policy

Policy elements		National open data policy of the Netherlands
Stage 2: policy content	Policy strategy and principles	Open by default (Ministerie van Binnenlandse Zaken en Koninkrijksrelaties, 2016)
	Actors involved in opening data	Governmental organizations collecting and creating data, IT providers
	Targeted open data users	Anyone, but particularly citizens and entrepreneurs
	Types of data opened and not opened	Data opened concerning many different topics (e.g. administration, culture and recreation, economy, finance, housing, international, agriculture, migration and integration (see data.overheid.nl)) Data not opened: (privacy) sensitive data, data that is not appropriate for opening.
	Policy measures and instruments	Three focus areas: (1) incentivisation and disclosure of datasets, (2) progress monitoring and quality, (3) supporting the disclosure, technology and users (Ministerie van Binnenlandse Zaken en Koninkrijksrelaties, 2016)
	Provision of (technical) support for opening data	Support for questions concerning technology, organization and licenses (Ministerie van Binnenlandse Zaken en Koninkrijksrelaties, 2016)
	Provision of (technical) support for open data use	User group to discuss operational and user barriers (Ministerie van Binnenlandse Zaken en Koninkrijksrelaties, 2016)
	Type of engagement of and interaction between data providers and users	Meetings between open data programme employees, data providers and data users, e-mail and data request forms (Data.overheid.nl, 2017a).
	Promotion of data and metadata	Promotion through social media, hackathons, user group meetings
	Data processing before opening	Open data should be provided as raw as possible (Data.overheid.nl, 2017c)
	Data quality	The organization owning the dataset is responsible for data quality aspects when opening and maintaining the data (Ministerie van Binnenlandse Zaken en Koninkrijksrelaties, 2016).
	Selected open data license and use conditions	Various licenses used (Algemene Rekenkamer, 2016)
	Data and metadata provision	Data offered through national data portal. Possible to search data sets, download data sets, CKAN API accessible for data uploading and downloading (Data.overheid.nl, 2017b), possibility to give feedback, not possible to contribute to the data portal directly (European Data Portal, 2016a).
	Numbers or percentages of opened datasets	11,676 datasets available (September 2017). Out of these datasets, 38% is provided by Statistics Netherlands and 43% is provided by the National Geo Register.
	Data access and availability (e.g. required registration, portal)	Data offered through various portals, often duplicated. Registration or login is usually not required.
	Way of presenting data and metadata to users (e.g. formats, standards)	National portal realized using CKAN. Various (inter)national metadata standards used, including OWMC (derived from DC) (Standaarden.overheid.nl, 2017) and DCAT-AP-NL (World Wide Web Consortium, 2014).
	Data update frequency	Differs per data provider and portal

Technical support is available for governmental organizations wishing to open up their data through the national open data portal data.overheid.nl. Support is given for questions concerning technology, organization and licenses (Ministerie van Binnenlandse Zaken en Koninkrijksrelaties, 2016). A user group has been set up to provide feedback to the national open data portal. The user group is open and meets several times per year (Ministerie van Binnenlandse Zaken en Koninkrijksrelaties, 2016). The ministry of the Interior and Kingdom Relations organizes several events each year, including meetings with users ('gebruikersgroep bijeenkomsten') several times per year. In addition, interaction is possible through e-mail and by filling out a data request form on the national open data portal (Data.overheid.nl, 2017a). The user of governmental data is promoted through Twitter, hackathons and user group meetings. Tweets about the Dutch open data policy appear frequently, and hackathons and user group meetings are organized several times per year. Hackathon are usually thematic, focusing on, for example, climate data, agriculture data or road infrastructure data.

According to the guideline of the Dutch federal government, open data should be open, without payment, available "as-is", free of rights, accessible without registration, computer processable, provided with metadata, complete, as raw as possible, timely and findable (Data.overheid.nl, 2017c). The organization owning the dataset is responsible for data quality aspects when opening and maintaining the data. This includes timeliness, accuracy, completeness, topicality and consistency of the data (Ministerie van Binnenlandse Zaken en Koninkrijksrelaties, 2016). In 2015, the Second Chamber called for more attention for data quality aspects (Second Chamber, 2015). The license used differs per dataset. Datasets are published both with open and more restricted licenses (Ministerie van Binnenlandse Zaken en Koninkrijksrelaties, 2016).

Data is offered through national data portal. It is possible to search and download data sets, and a CKAN API is accessible for data uploading and downloading (Data.overheid.nl, 2017b). It is possible to give feedback regarding datasets or the data portal, but it is not possible for data providers or users to contribute to the data portal directly (European Data Portal, 2016a). In total, 11,676 datasets were available in September 2017. Out of these datasets, 38% is provided by Statistics Netherlands and 43% is provided by the National Geo Register. Data is not only offered through the national open data portal, but also through other portals, resulting in fragmentation. Registration is usually not required, in line with the national guidelines.

The national open data portal has been realized using the Comprehensive Knowledge Archive Network (CKAN), which is also used for open data portals in several other European countries. The national portal offers data in many different formats (see Algemene Rekenkamer (2016) for an overview). Various (inter)national standards are used to present the data on the national data portal, including:

- Overheid.nl Web Metadata Standaard (OWMS). This national standard is derived from the international Dublin Core (DC) standard (Standaarden.overheid.nl, 2017).
- Data Catalog Vocabulary (DCAT). This international standard allows for the exchange of datasets between data registers (World Wide Web Consortium, 2014).

Table 3.4 Policy environment characteristics of the Dutch open data policy

Policy elements		National open data policy of the Netherlands
Stage 3: Policy implementation	Performance indicators concerning open data provision (e.g. number of datasets opened, machine-readability of data)	Performance for open data provision is measured in various ways, e.g.: The number of opened datasets compared to the number of available datasets (Ministerie van Binnenlandse Zaken en Koninkrijksrelaties, 2017b) The opening of municipal high-value datasets (Ministerie van Binnenlandse Zaken en Koninkrijksrelaties, 2017a) The scores in international benchmarks: the Open Data Barometer (World Wide Web Foundation, 2016), the European Open Data Benchmark (European Data Portal, 2016b) and the Global Open Data Index (Open Knowledge International, 2016) Research by the National Audit office (Algemene Rekenkamer, 2015, 2016).
	Performance indicators concerning open data use (e.g. the number of data users, number of dataset downloads, type of data use)	Performance for open data provision is measured mainly by.: The scores in international benchmarks: the Open Data Barometer (World Wide Web Foundation, 2016), the European Open Data Benchmark (European Data Portal, 2016b) and the Global Open Data Index (Open Knowledge International, 2016) Research by the National Audit office (Algemene Rekenkamer, 2015, 2016).

The data update frequency differs per dataset and data provider (Table 3.4).

The performance of the Netherlands in opening its data is measured in various ways. First, since 2015, an annual 'data inventory' is carried out, aimed at identifying all available datasets within governmental organizations and at examining which datasets are appropriate for opening. An inventory template has been developed and the inventory process is open and available as open data. An inventory is made for ministries, municipalities, provinces and district water boards. The number of opened datasets is compared to the number of available datasets (Ministerie van Binnenlandse Zaken en Koninkrijksrelaties, 2017b). The results of the inventory are reported on a dedicated website (https://data.overheid.nl/data-inventarisatie) and in a letter to the Second Chamber (Minister of the Interior and Kingdom Relations, 2017). Second, municipal high value datasets have been identified (Ministerie van Binnenlandse Zaken en Koninkrijksrelaties, 2017a). The list of high value datasets should help municipalities in prioritizing the opening of certain datasets. Third, the progress in opening data and use is monitored by examining the scores of international benchmarks: the Open Data Barometer (World Wide Web Foundation, 2016), the European Open Data Benchmark (European Data Portal, 2016b) and the Global Open Data Index (Open Knowledge International, 2016). In addition, the National

Table 3.5 Policy evaluation characteristics of the Dutch open data policy

Policy elements		National open data policy of the Netherlands
Stage 4: evaluation: public value realized?	Political and social value (e.g. increased transparency)	Scores in international benchmarks: The Open Data Barometer, ranked #8 (World Wide Web Foundation, 2016) The European Open Data Benchmark, the Netherlands is viewed as a 'trendsetter' (European Data Portal, 2016a) The Global Open Data Index, ranked #20 (Open Knowledge International, 2016)
	Economic value (e.g. economic growth)	Unknown
	Technical and operational value (e.g. ability to reuse data)	Data inventory findings are available No monitoring of the opening of municipal high value datasets so far Many missed opportunities, many more datasets can be opened (Algemene Rekenkamer, 2015, 2016).

Audit Office examines the Dutch open data progress (Algemene Rekenkamer, 2015, 2016) (Table 3.5).

Regarding the political and social value, the scores of international benchmarks are reported. The Netherlands is ranked place 8 in the Open Data Barometer (World Wide Web Foundation, 2016). Out of the maximum score of 100 points, the economic impact receives a score of 47, the political impact a score of 63 and the social impact a score of 50 (World Wide Web Foundation, 2016). According to the European Open Data Benchmark, the Netherlands can be viewed as a 'trendsetter', together with countries like the United Kingdom, France and Finland (European Data Portal, 2016a). The Netherlands is ranked place 20 in the Global Open Data Index (Open Knowledge International, 2016). The Dutch open data policy has a score of 54% out of the maximum score of 100%. 40% of the defined data types is open as defined by the Open Definition (Open Knowledge International, 2016). At the same time one should keep in mind that each benchmark uses different indicators and each of them has its advantages and disadvantages (Susha et al., 2015).

Information regarding the created economic value is lacking. As far as the technical and operational value are concerned, the number of available datasets compared to the number of opened datasets is reported at https://data.overheid.nl/rijksbrede-inventarisatie-2017. It is also reported how many datasets cannot be opened because of, for instance, privacy concerns and how many datasets are still under investigation. The National Audit Office states that there are many missed opportunities (Algemene Rekenkamer, 2015, 2016). Not so many new datasets have been opened recently (only datasets already available at other portals have been copied to the national portal), whereas many datasets can still be opened (Algemene Rekenkamer, 2016). There is also no process of monitoring the opening of municipal high value datasets in place at the moment, although the high value list has only been created in 2016 (Table 3.6).

Table 3.6 Policy change/termination characteristics of the Dutch open data policy

Policy elements	National open data policy of the Netherlands
Stage 5: policy change or termination	Gradual development of open data policy. Several policy documents have been developed. Ministerie van Binnenlandse Zaken en Koninkrijksrelaties (2017c) provides an overview.

The Dutch national open policy is in place for several years now and improvements are gradually being made. So far, the policy has not been changed considerably, yet it has been made more specific and detailed (e.g. by adding more specific overviews of how many datasets are available through the data inventories), and it has expanded (e.g. by also providing data of more municipalities and provinces through the national open data portal and by connecting to Statistics Netherlands and the national Geo Register). Open data will remain an important focus area for the Dutch government in the following years, as indicated by the government that was formed in 2017: "The government own considerable general, public information. This data will be made findable and accessible in the form of open data" (Bureau Woordvoering Kabinetsformatie, 2017, p. 7).

3.7 Conclusions and Lessons Learned Concerning Open Data Policies

In this chapter we looked into open data directives and policies. Directives promote the development of open data policies and provide a high-level framework. We provided examples of elements of directives and policies, we discussed existing open data directives and policies, we provided an example of the elements of the Dutch national open data policy, and we discussed lessons learned from open data policy development. This chapter provided us with various lessons that can be learned concerning open data policies in general. First, several frameworks for comparing open data policies have already been developed, and they show that a wide variety of open data policies exist. Existing policies have a different focus and open data policies may encompass different elements. The elements of open data policies that we described in this chapter are not covered by every policy. There is variety in the policy environment and context, the policy content (the input), the performance indicators (the output), the attained public values (the impact) and policy change or termination (the feedback). The differences between open data policies may indicate that open data policies stimulate the provision and use of open data in different ways, and this could reveal opportunities for learning from each other (Zuiderwijk & Janssen, 2014a).

Open data policies may not only include statements in documents, but also the actual behaviour and practice of governments. Often this is overlooked. Open data policies should not only focus on the opening of data, but they should pay special attention to improving the use of and value creation with open data. Open data policies have been developed all over the world, both in developed and in developing countries (Nugroho, Zuiderwijk, Janssen, & de Jong, 2015) and at different administrative levels (international, national, state, regional, local – see Table 3.1). There is no best policy, as open data policies depend on the context in which they are created and on the policy objectives.

Open data policies can also be criticized for several reasons. As an example, open data policies are usually formulated on a high level of abstraction. They are often not very specific, since they also need to leave enough freedom for interpretation and application, which can make it difficult for those who need to implement the policy to use the policy as a guideline. Another example is that the user perspective is often lacking in open data policies. Open data policies are usually focused on what governments aim to achieve and how they want to do this, but they often lack the mechanisms that are required to identify and address the need of open data users, although more the user perspective is being acknowledge increasingly.

Moreover, having a policy in place does not necessarily mean that this policy will be implemented. Policy makers need to be aware that merely the design of open data policies is not enough, and additional measures are required. For example, governmental agencies may not be motivated to open up governmental data or they may not have the necessary resources to do so, which could lead them to ignore the design policies. It is also possible that government agencies that collect and hold data are not aware of the develop policies and the requirement to open up their data, or they may not know how to design processes required for opening up data within their organization. Open data is a quickly developing field that is influenced by developments in related fields, such as EU General Data Protection Legislation (GDPR). New legislation may make government agencies reluctant to open up their data, since it may not yet be clear how the new legislation should be interpreted in context of their organization. A lack of stability and reliability of legal frameworks are not only likely to lead to less opening up of governmental data, but in combination with other barriers (e.g. the low quality of data released) they are also likely to lead to less open data usage.

Chapter 4
Organizational Issues: How to Open Up Government Data?

When publishing data, governmental organizations are often hindered by issues such as the lack of standard procedures, the threat of privacy violations when releasing data, the risk of accidentally releasing policy sensitive data, the risk of data misuse, and problems with data ownership.

4.1 Introduction

Governments create and collect enormous amounts of data, for instance concerning voting results, transport, energy, education, and employment. These datasets are often stored in an archive that is not accessible for others than the organization's employees. To attain benefits such as transparency, engagement, and innovation, many governmental organizations are now also providing public access to this data. However, in opening up their data, these organizations face many issues, including the lack of standard procedures, the threat of privacy violations when releasing data, accidentally releasing policy-sensitive data, the risk of data misuse, challenges regarding the ownership of data and required changes at different organizational layers. These issues often hinder the easy publication of government data.

In Chap. 2 we already discussed the open data lifecycle, including the steps that organizations take in opening data. This chapter discusses these steps and their related issues and potential effects more in depth. In this chapter we first discuss issues that governmental organizations face when opening up their data. We give an overview of all the issues, including the potential positive and negative effects, and then discuss each of them in detail, with a related example from the open government domain. Subsequently, we provide a use case that describes solutions to overcome some of the outlined issues. Thereafter, we describe best practices that function as guidelines for governmental organizations that want to open up their data. Such guidelines can be used by public organizations to improve their open data publishing processes. Ultimately, the implementation of the guidelines reduces barriers, stimulates the publication of government data, and contributes to attaining

Y. Charalabidis et al., *The World of Open Data*, Public Administration and Information Technology 28, https://doi.org/10.1007/978-3-319-90850-2_4

the benefits of open data. Discussions with practitioners showed that the guidelines could improve the open data publication process.

4.2 Organizational Issues for Opening Up Government Data

Let us imagine that you are a civil servant working for a governmental organization, for instance, a ministry. As part of your daily tasks at the ministry, you have collected a number of datasets, and you consider opening the collected data. Which aspects do you need to consider? The main issues that public organizations may face when opening up their data are depicted in Table 4.1 (adapted from Janssen, Charalabidis, & Zuiderwijk, 2012; Susha, Zuiderwijk, Charalabidis, Parycek, & Janssen, 2015; Zuiderwijk, 2015a; Zuiderwijk & Janssen, 2015; Zuiderwijk et al., 2012b). We provide an example of each organizational issue and explain these issues further in the following sub sections.

Table 4.1 Organizational issues for opening up government data

Types of organizational issues	Organizational issues	Example
Data related issues	Potential privacy breaches	The Ministry of Justice collects data concerning crime victims and offenders. The data may be of interest to the public, yet it can only be opened up after it has been anonymized and/or aggregated.
	Data sensitivity and security	Data collected by the Ministry of Education may be sensitive, since it contrasts information provided by the responsible Minister of Education.
	Embargo period	A researcher working at a ministry first wants to publish an article and a report using the collected data. The data can only be opened after the article and report have been published.
	Data openness, lack of control over its use and lack of trust in the data user	A dataset concerning employment has been published online. After publication, the dataset is copied into various online repositories. Although this enhances openness, it is not clear to the data provider anymore at which places the data is available and how it is used. The data may be misused.
	Data quality	Some datasets are of high quality (i.e., they are complete, accurate, timely, and reliable), whereas for some datasets, the quality is low (e.g., the dataset is not complete) or it is unknown what the quality level is.
	Data documentation	Interesting domain-specific data has been collected by a government official, yet the metadata describing the data is very limited and not sufficient for an outsider to make sense of the data.

(continued)

Table 4.1 (continued)

Types of organizational issues	Organizational issues	Example
Infrastructure and process-related issues	Lacking infrastructure and resources (including skills and training)	A municipality wants to become more transparent and show the municipality's inhabitants which data it collects, yet the municipality does not have the human and technical resources and infrastructure to make the data available to the public.
	Unclear or shared ownership	Two governmental organizations have worked together and integrated their data registers and datasets to obtain new insights. They share the ownership of the newly created dataset, but they disagree about opening the data.
	Changes to organizational processes required	A governmental organization willing to open data by default needs to change not only the data opening processes, but also the processes that precede the opening (e.g., during the data collection processes), since considerable metadata need to be collected simultaneously alongside the data itself. Changing work processes is complicated and may require additional work for several employees, whereas there are no direct incentives for them to change their work processes.
	Negative consequences for the government	Gas drillings in the Netherlands create large financial benefits for the government. Open data about earthquakes was used by lobbyists to demonstrate against the gas drillings that caused earthquakes in the northern part of the Netherlands. Under pressure, the Dutch government had to decide to reduce the amount of gas derived from this part of the Netherlands. Thus, the publication of government data resulted in less income from gas drillings.
	Benefits obtained by others than the government	The Ministry of Environment and Infrastructure puts much effort into opening datasets concerning traffic, road conditions, license plates and vehicle information. A company uses this data and creates an application that presents the information through a user-friendly interface that citizens need to pay for. The company creates revenue out of selling the application, whereas the government does not.

Adapted from Janssen et al. (2012), Susha et al. (2015), Zuiderwijk (2015a), Zuiderwijk and Janssen (2015), Zuiderwijk et al. (2012b)

4.2.1 Data-Related Issues

4.2.1.1 Potential Privacy Breaches

An important issue for governmental organizations opening data concerns the risk to violate individuals' privacy (Kalidien, Choenni, & Meijer, 2010; Kulk & van Loenen, 2012). Regardless of the amount of effort put into removing privacy sensitive content from datasets, privacy cannot be guaranteed. Even if an individual dataset does not violate a person's privacy, the combination of multiple datasets or the combination of open datasets with information from the media may allow for identifying persons in a dataset (Zuiderwijk & Janssen, 2014b), especially when open data is combined with social media data (Nieuwenhuijs, 2014). For instance, let us imagine that a researcher locates two datasets. The first dataset contains data about the number of crime offenders in a certain neighbourhood per type of crime (e.g., sex offences). With this dataset, someone can identify in which neighbourhood sex offenders live. The second dataset reveals the number of crime offenders per type of crime and per gender and age category. On themselves, these datasets do not allow identifying a particular person. However, their combination may allow this. If there is only one female sex offender in the age category of 70 years and older in a certain neighbourhood, identification of the particular offender becomes possible. With additional information from the media, the person might be identified. If one organisation releases the first dataset from the example and another organisation releases the second dataset, the privacy of citizens can easily be violated (example adapted from Kalidien et al., 2010).

Data protection legislation often prescribes on a very general level how one should handle privacy sensitive data, and thus it does not give much guidance for removing (privacy) sensitive information from datasets (Zuiderwijk & Janssen, 2014b). Laws and regulations need to give sufficient space for the interpretation of privacy sensitivity and therefore they cannot be too specific (idem). Furthermore, the situation in different countries might vary, as privacy is valued more in some countries than in others (idem). In sum, guidelines about privacy sensitivity partly help to identify which data cannot be published, yet much interpretation effort by the data provider is still required, and combining data could still lead to identifying a person or company (Zuiderwijk & Janssen, 2014b). When privacy-sensitive data is opened, this can result in considerable negative attention and might lead to reputation damage of the organization that opened the data or might lead to a decrease of trust in the government in general.

4.2.1.2 Data Sensitivity and Security

In addition to privacy-sensitive data, some governmental datasets are sensitive in other ways. For instance, data can be policy-sensitive. Whereas privacy-sensitive data refers to violating privacy of an individual or company, policy-sensitive data refers to data that may have negative consequences for government officials, responsible for a policy or for politicians working on issues related to these datasets. The data may contrast certain statements or positions, posited by a politician or it may show that a certain policy proposed by an important politician does not work as expected (Zuiderwijk, Janssen, Choenni, & Meijer, 2014). Governmental data may also be sensitive in the sense that it contains information that is considered a state secret and should not be provided to politicians of other countries, as it may block negotiation processes, or it may negatively influence ongoing alliances.

Sensitive data is often not released. Data sensitivity is an issue for organizations aiming to open up government data. On the one hand, these organizations are willing to become more open, yet on the other hand, determining whether a dataset is sensitive is complicated and accidentally releasing sensitive data could have many undesired consequences (Zuiderwijk et al., 2014). For example, opening sensitive data could damage the reputation of an individual (including politicians) or organization, it could also be dangerous, or lead to the resigning of a minister or conflicts with other countries.

Determining which data is sensitive and which data is not requires an examination of each individual dataset that an organization considers opening, also bearing in mind the context of to whom the data will be opened and with which other data the data might be released and potentially combined. This consideration requires interpretation by a human being, and mistakes might be made (Zuiderwijk, 2016). Since sensitive data is often not released, the data that is released usually favors policies set and arguments provided by politicians in place. Data that might demonstrate the opposite and give a different perspective might not be opened (Zuiderwijk & Janssen, 2014b).

4.2.1.3 Embargo Period

For each governmental dataset that is considered to be opened, a government official needs to ask the question of whether there are reasons for not yet opening the data, which may include an embargo period, i.e. a period in which a dataset is not publishable, although it might become publishable in the future. Some datasets may be publishable, but just not immediately after they have been collected. Reasons for having an embargo period are diverse. As an example, civil servants may first use the collected data to write a governmental report, and the opening of the data should be delayed until politicians presented and discussed the report (e.g., at the level of the national government) (Zuiderwijk & Janssen, 2014b). Some reports need to remain confidential, and thus the data also remains closed. A second reason for setting an embargo period is that civil servants may want to write an article (e.g., a scientific article) based on these data (idem). Data publication then has to be postponed, until the article has been published, which can take years. Other reasons for an embargo period include that the governmental organization that collected the data may want to conduct follow-up research using the data (idem), or that the data is too sensitive at a certain moment (e.g. when the national government is discussing a certain issue or topic and developing policies and legislation in this area). This data might become less sensitive over time.

Embargo periods have several advantages. Some datasets may still be opened when an embargo period is used, whereas they would not have been opened otherwise. Embargo periods give governmental organizations time to think data release through and may prevent wrongfully publishing data. It also allows for still publishing data that has become less sensitive over time. Embargo periods also have disadvantages. Datasets may become less useful over time; their quality reduces as timeliness of the data reduces at the moment of data publication (Zuiderwijk, 2016; Zuiderwijk & Janssen, 2014b).

4.2.1.4 Data Openness, Lack of Control Over Its Use and Lack of Trust in the Data User

To which extent should openness be provided? In one respect releasing governmental data may provide the public with more insight in what governmental processes encompass and what public agencies do. Datasets may be copied to many repositories and become available to a large audience. In another respect, opening governmental data to the public may result in too much openness. When datasets are open and become available at different places, this does not only enhance openness, but this also makes it difficult for the data provider to keep track of where the data is available and how it is used. The data provider may fear misuse of the data and may not completely trust potential data users. In addition, public agencies may accidentally release sensitive data that should not have been released. This may result in a more negative image of the government and may decrease the public's trust in the government.

4.2.1.5 Data Quality

Another consideration when opening governmental data concerns the quality of the data. Important data quality dimensions include completeness, timeliness, accuracy and consistency (Batini, Cappiello, Francalanci, & Maurino, 2009). Civil servants may decide to disclose data without having insight in its quality. Consequently, they may publish data that is incomplete, inaccurate, invalid, or unreliable. This may lead to low value and exploitation possibilities and thus to low reusability (also see Chap. 7 concerning value creation). Low quality data may also be published on purpose where publishing low quality data is considered a "quick win". Proponents for releasing and opening low data quality data argue that the release of low quality data could help in identifying the dimensions on which the quality of the data is poor, so that governmental data providers can improve these dimensions (see Chap. 7). The crowd can comment on the data and can try to improve low-quality data. Feedback to data providers regarding data quality might create incentives for the data publisher to improve the data (Zuiderwijk & Janssen, 2014b).

At the same time, some data users may not notice that the data is of poor quality. The low-quality data may be reused, and decisions and conclusions may be based on this data. This may result in wrongful decisions and little value creation. A dataset with many missing values or variables may be misinterpreted or may not be useful at all. Opponents of releasing low-quality data state that datasets need to have at least a certain level of quality before they can be published (Zuiderwijk, 2016) and should be in a format that enables reusability (also see Chap. 5 concerning interoperability). Both the arguments of the proponents and the opponents can be valid and assessing whether low-quality data can be opened requires a trade-off per dataset (Zuiderwijk & Janssen, 2014b). Data quality can also be subject of evaluation (see Chap. 8).

4.2.1.6 Data Documentation

Another consideration for releasing governmental data concerns data documentation. To be able to use open government data, users need to have information about the meaning of the data and the semantics need to be clear. They need data documentation to understand how the data can be used. For instance, to be able to find a book in the library, a person needs to know in which category he or she should look for the book. Part of the collected governmental data is poorly-documented and might be misinterpreted if it would be opened. Data may concern a specific domain (e.g. earth observations or the criminal justice chain) whereas data users do not necessarily have the domain-specific knowledge that is required to interpret the data correctly. This could lead to incorrect conclusions derived from data analysis results (Zuiderwijk & Janssen, 2014b). Considerable documentation is then required to understand the data. At the same time, adding considerable documentation to governmental datasets requires effort and time investments from the data provider, since this information often cannot be derived automatically from the data provider's systems (Zuiderwijk, 2016).

4.2.2 Infrastructure and Process-Related Issues

4.2.2.1 Lacking Infrastructure and Resources

Opening data requires the availability of an infrastructure. An Open Government Data Infrastructure can be defined as "*a shared, (quasi-)public, evolving system, consisting of a collection of interconnected social elements (e.g. user operations) and technical elements (e.g. open data analysis tools and technologies, open data services) which jointly allow for OGD use.*" (Zuiderwijk, 2015a, p. 45). Open data infrastructures are shared by a variety of actors and systems. Actors, such as governments, researchers, and citizens, can use the infrastructure, for example, by downloading and processing a dataset. Open data infrastructures consist of technical elements, such as tools and technologies (e.g., tools and platforms to analyze open data), and social elements, such as user operations and interactions (e.g., communication from the provider to the user about how the infrastructure can be used) (Zuiderwijk, 2017). Data, platforms and people are connected through the open data infrastructure (idem). Data, information, and knowledge are important resources that are transferred and exchanged in open data infrastructures. Such infrastructures evolve through the development of new technologies and through the adaptation of the infrastructure by people. All infrastructure elements are needed in combination to ensure that the infrastructure can function. The lacking or malfunctioning of one element results in problems for the functioning of the entire infrastructure. For example, if data providers and users are not connected, or if platforms are lacking of functionality and components, it becomes difficult to find and use the data and attain the potential benefits. In practice, open data infrastructures are still under development and various challenges need to be overcome. For instance, many open data infrastructures are mainly focused on the opening of governmental data and less on the use of the data, whereas the data use should eventually lead to attaining the benefits.

Opening data also requires resources of governmental organizations. Human resources are needed, such as computer skills, skills concerning data interpretation (to assess whether a dataset can be opened), resources for uploading datasets (e.g., time and effort), and resources related to the selection of tools for opening and sharing data. Data opening also requires technical resources, such as an internet connection and tools for processing and viewing datasets, as well as information and data resources, such as a repository of open data sets. Civil servants may need to be trained to develop the skills needed to open up governmental data.

4.2.2.2 Unclear or Shared Ownership

Data opening requires an assessment of ownership of the data. Often datasets are created through a collaboration of multiple people and organizations, and it may be unclear who owns the data, or involved parties may disagree about whether a dataset can be opened. Even if the collaborators agree on opening datasets that they created together, a potential risk is that it may be unclear who is responsible and accountable if something goes wrong, for instance, if data is misused. Datasets owned by organizations from different countries may also have to comply with different laws and policies concerning data protection (Faerman, McCaffrey, & Slyke, 2001; Zuiderwijk et al., 2014).

4.2.2.3 Changes to Organizational Processes Required

To really become open and systematically publish open datasets, governmental organizations need to make changes at different organizational layers (Van Veenstra & van den Broek, 2013) and for many organizations it is unclear how the publishing process could be modified to improve it and to institutionalize data opening

(Zuiderwijk & Janssen, 2013; Zuiderwijk et al., 2014). The open data literature is more focused on the development of open data portals and infrastructure, data publication, functionality and other instruments to release and use open data. Although this is an important first step, it is important to transform the structure of organizations and change the cultures and incentives to open data so that structural changes are made and so that opening data becomes part of the daily work processes, routines, and procedures (Zuiderwijk & Janssen, 2014b).

4.2.2.4 Negative Consequences for the Government

Releasing governmental data does not only have the potential to result in benefits, but can also lead to negative consequences for the government. Several scholars mention that opening data may result in, for example, the benefit of transparency (e.g., Bertot, Jaeger, & Grimes, 2010; Böhm et al., 2012a), yet transparency may also result in a more negative image of the government. If datasets of low quality are opened, or if opened datasets reveal the misbehavior of civil servants, this might decrease trust in the government (Zuiderwijk & Janssen, 2014b). Furthermore, opened datasets may be misused or misinterpreted (Kalidien et al., 2010; Kulk & van Loenen, 2012; Zuiderwijk et al., 2014).

4.2.2.5 Benefits Obtained by Others Than the Government

One of the challenging aspects of the open data process is that governmental organizations invest resources by opening data, whereas others benefit from this. The data providers are often not the ones who benefit, although they spend time and effort on opening the data. Policy makers working for governmental organizations may be able to use insights that data users outside the government obtained from the analysis of the governmental data. This may concern, for example, policy-making in the area of social security, economy, justice, elections, health, energy, and transport (Zuiderwijk, 2015a). Zuiderwijk (2015a, p. 4) describes the example of governmental policy-makers, who use insights obtained from the use of open crime data by non-governmental researchers to develop governmental policies about security measures and police surveillance. However, often users and (governmental) policy-makers do not communicate about the results of open data use and what lessons can be learned from this (Zuiderwijk, 2015a).

4.3 Use-Case: Solutions to Overcome the Issues

In this section, we discuss two use-cases that contain solutions on how to overcome some of the above-mentioned issues. They focus particularly on the risk of privacy violation (from an administrative perspective), and on the issue that benefits are usually obtained by others than the governmental organization that is opening the data (from a research perspective).

4.3.1 Solutions to Reduce the Risk of Privacy Violation (Administration View)

Yin (2017) provides an overview of solutions to enhance privacy and to reduce the risk on privacy violation for information sharing in general. He states that such solutions should combine technical and governance or managerial aspects. One category of technical aspects is referred to as Privacy Enhancing Technologies (PETs), including tools for encryption, policy, filtering, and anonymization (Yin, 2017). More examples concerning these PETs can be found in Seničar, Jerman-Blažič, and Klobučar (2003). The governance or managerial aspects mentioned by Yin (2017) include the development of legislation for data protection, self-regulation (voluntary privacy protection mechanisms) and privacy by design (building in privacy upfront). Privacy by design can be defined as an approach to protect privacy by embedding it into the design specifications of technologies, business practices, and physical infrastructures (Cavoukian, 2011).

In addition, Ali- Eldin, Zuiderwijk, and Janssen (2017) developed a model for privacy risk scoring for open data. The model consists of open data attributes and privacy risk mitigation measures. The open data attributes influence the decision of whether or not to open up a dataset. They include the need for openness, the criticality/importance level, the level of cyber security threat, the trustworthiness of the data provider, and the restrictions of use (including the type of user, the physical location that the data is accessed from, and the purpose the data is used for). Each attribute has different values and each value has a different score. Adding up the scores results in a Privacy Risk Indicator (low, low-medium, medium-high, or high). Based on the indicator level, a Privacy Risk Mitigation Measure (PRMM) is proposed. If the PRI is low, only the removal of identifiers from a dataset is proposed, using tools such as Anonymizer, ARX, or Camouflage's-CX-Mask. If the IRP is on the low-medium level, the model recommends altering quasi-identifiers to reduce identity leakage. "Quasi-identifiers are data types which if linked with other datasets can reveal real identities" (p. 150). If the IRP indicates medium-high privacy risks, the model suggests removing sensitive items, and when the IRP is high, it is advised not to publish the data at all. Each defined privacy risk mitigation measure should be applied before publishing a government dataset on the internet (Ali-Eldin et al., 2017).

These are just a few examples of data protection solutions, but more of them exist. Furthermore, each of the provided solutions also has its drawbacks. For instance, anonymization is often not sufficient, as the combinations of datasets could still lead to re-identifying persons and their activities.

4.3.2 *Solutions to Develop an Open Data Infrastructure That Enhances the Coordination Between Open Data Actors (Research View)*

In practice, benefits of open data are usually obtained by others than the governmental organization that is opening the data. Zuiderwijk (2015a) argues that the use of open government can support open data publication and governmental policy-making, since governmental open data providers and governmental policy makers can learn from the insights obtained using open data. This is challenging, since this requires several actors – dependent on each other – to work together and to coordinate their activities. Zuiderwijk (2015a) proposes the design of an open data infrastructure to enhance the coordination of open data use by researchers. An infrastructure for Open Government Data (OGD) is defined as a shared, (quasi-) public, evolving system, consisting of a collection of interconnected social elements (e.g., user operations) and technical elements (e.g., open data analysis tools and technologies, open data services) which jointly allow for OGD use (p. 269). The theory focuses on the coordination of searching for and finding OGD, OGD analysis, OGD visualization, interaction about OGD, and OGD quality analysis. "*In the context of this study, three design propositions were elicited:*

- Metadata positively influence the ease and speed of searching for and finding OGD, OGD analysis, OGD visualisation, interaction about OGD and OGD quality analysis.
- Interaction mechanisms positively influence the ease and speed of interaction about OGD.
- Data quality indicators positively influence the ease and speed of OGD quality analysis." (Zuiderwijk, 2015a, p. 270)

The metadata model, the interaction mechanisms, and the data quality indicators need to be combined to support searching for and finding OGD, OGD analysis, OGD visualisation, interaction about OGD, and OGD quality analysis. Building on 22 coordination design principles, 40 metadata design principles, 15 interaction design principles, and 4 data quality design principles, the system design, the coordination patterns and the function design of the OGD infrastructure were developed. Evaluations of a prototype, integrating the designed infrastructure, provided support for the three propositions (Zuiderwijk, 2015a).

4.4 Best Practices

The Share-PSI 2.0 project has created an overview of best practices for sharing open government data (Share-PSI 2.0, 2016a), as depicted in Table 4.2. One of the main aims of the Share PSI 2.0 best practices is the Implementation of the (revised) PSI Directive (European Commission, 2003, 2013c).

Table 4.2 Best practices for sharing open government data

Best practice (Share-PSI 2.0, 2016a)	Description (Share-PSI 2.0, 2016a)
Categorise openness of data	Public sector organizations can create a system in which the openness of data is categorized, so that it becomes easier for them to determine with whom data can be shared.
Dataset criteria	Public sector organizations can prioritize the publication of some datasets in comparison to others. For example, datasets that contribute to transparency, datasets that help with cost reductions, or highly structured datasets may be published first.
Develop an Open Data publication plan	Public sector organizations are recommended to develop a plan in which they address the abovementioned issues and determine which datasets are fit for publication as open data, which requirements the internal and external stakeholders have, and which potential benefits, risks and costs of data opening play a role.
Develop and implement a cross agency strategy	In addition to a plan for individual organizations, it is recommended to develop and implement an open data strategy that coordinates the efforts of multiple organizations. In most of the EU countries these strategies have been interpreted in guides for publishing data across agencies and in some cases, they are incorporated in the national law through presidential or ministerial decrees. The strategy should also foresee the way the opening will be implemented by the public-sector organisations. An example could be a stage strategy focusing at the first level in a quick win publishing data as quickly as possible before a specific deadline and at the second level focusing on the quality improvement (Share-PSI 2.0, 2016b).
Enable feedback channels for improving the quality of existing government data	The quality of governmental data can be improved by facilitating feedback channels for users to report errors, inconsistencies, and incompleteness in openly available datasets.
Enable quality assessment of open data	Since data quality is considered to be subjective, depending on the context, data quality should be measured in different ways all along the data pipeline (not only at the front end). These measures should sustainably raise data quality.
Encourage crowdsourcing around PSI	The open data community can help to improve the quality and quantity of available datasets and can enthuse potential data users.
Establish an Open Data ecosystem	An open data ecosystem can enable the uptake of government data and information for reuse, so that services can be built for citizens.
Establish Open Government Portal for data sharing	Government data should be published through open data portals that provide potential users with easy access to a searchable hub for multiple datasets.
High level support	Senior staff should support open data actions.
Holistic metrics	Value generation using open government data and costs of making this data available have to be assessed in respect to large-scale detour effects and not only at the level of the data providing agency.

(continued)

Table 4.2 (continued)

Best practice (Share-PSI 2.0, 2016a)	Description (Share-PSI 2.0, 2016a)
Identify what you already publish	To make it easier to decide what data should be made available, it is useful to examine which datasets are already opened. An inventory must be created and maintained of already opened data.
Open Data business models and value disciplines	A business model should be described, explaining how value is created and captured for data opened by a certain public organization (at all levels) and what the expected results are.
Open up public transport data	Transport data (e.g. timetables, service disruptions and accessibility) is considered as high-value data and can be used to create a better experience for transport users, greener cities by using collective transport, and more efficient companies.
Open up research data	Opening up research data promotes the discoverability and measurability of scientific achievements, and can stimulate innovation, economic growth and education.
Provide PSI at zero charge	The ability to use open data without payment unlocks maximum commercial and non-commercial potential.
Publish overview of managed data	Public organizations must publish an overview of datasets that it manages, so that potential users know what may be(come) available.
Publish statistical data in Linked Data format	The Linked Data format is an approach for expressing data in a standardised machine-readable manner and for providing a recommended set of metadata terms to describe the data.
(Re)use federated tools	Federated/distributed tools for open data collection can be used to automatically publish all the (meta)data published on the websites of each public entity. This can result in a global index of reusable open datasets.
Standards for Geospatial Data	For many public and privacy organizations location is essential and thus geospatial data should be shared in a way most likely to be re-usable: adhering to standards.
Support Open Data start ups	Open data provides a good basis for entrepreneurship, allowing for the development of added value services by citizens and small enterprises. Start-ups can be supported through the collaboration between universities (potential entrepreneurs), private and public funding organisations (chambers of commerce, municipalities, start-up investors) and experts (coaches and mentors).

Share-PSI 2.0 (2016a)

In addition, technical best practices related to the publication and usage of data on the Web have been developed by the World Wide Web Consortium (W3C) (World Wide Web Consortium, 2017). The best practices facilitate the interaction between data publishers and data users, and emphasizes that data should be discoverable and understandable by humans and machines. It also states that the use of data should be discoverable and that the efforts of the data publisher should be acknowledged and recognized (Table 4.3).

More information concerning each W3C Best Practice can be found at http://www.w3.org/TR/dwbp/.

Table 4.3 Technical best practices related to the publication and usage of data on the Web

	Best practice (World Wide Web Consortium, 2017)	Description (World Wide Web Consortium, 2017)
Metadata	1. Provide metadata	Provide metadata for both human users and computer applications.
	2. Provide descriptive metadata	Provide metadata that describes the overall features of datasets and distributions.
	3. Provide structural metadata	Provide metadata that describes the schema and internal structure of a distribution.
Data licenses	4. Provide data license information	Provide a link to or copy of the license agreement that controls use of the data.
Data provenance	5. Provide data provenance information	Provide complete information about the origins of the data and any changes you have made.
Data quality	6. Provide data quality information	Provide information about data quality and fitness for particular purposes.
Data versioning	7. Provide a version indicator	Assign and indicate a version number or date for each dataset.
	8. Provide version history	Provide a complete version history that explains the changes made in each version.
Data identifiers	9. Use persistent URIs as identifiers of datasets	Identify each dataset by a carefully chosen, persistent URI.
	10. Use persistent URIs as identifiers within datasets	Reuse other people's URIs as identifiers within datasets where possible.
	11. Assign URIs to dataset versions and series	Assign URIs to individual versions of datasets as well as to the overall series.
Data formats	12. Use machine-readable standardized data formats	Make data available in a machine-readable, standardized data format that is well suited to its intended or potential use.
	13. Use locale-neutral data representations	Use locale-neutral data structures and values, or, where that is not possible, provide metadata about the locale used by data values.
	14. Provide data in multiple formats	Make data available in multiple formats when more than one format suits its intended or potential use.
Data vocabularies	15. Reuse vocabularies, preferably standardized ones	Use terms from shared vocabularies, preferably standardized ones, to encode data and metadata.
	16. Choose the right formalization level	Opt for a level of formal semantics that fits both data and the most-likely applications.

(continued)

Table 4.3 (continued)

	Best practice (World Wide Web Consortium, 2017)	Description (World Wide Web Consortium, 2017)
Data access	17. Provide bulk download	Enable consumers to retrieve the full dataset with a single request.
	18. Provide Subsets for Large Datasets	If your dataset is large, enable users and applications to readily work with useful subsets of your data.
	19. Use content negotiation for serving data available in multiple formats	Use content negotiation in addition to file extensions for serving data available in multiple formats.
	20. Provide real-time access	When data is produced in real-time, make it available on the web in real-time or near real-time.
	21. Provide data up to date	Make data available in an up-to-date manner, and make the update frequency explicit.
	22. Provide an explanation for data that is not available	For data that is not available, provide an explanation about how the data can be accessed and who can access it.
Data access – APIs	23. Make data available through an API	Offer an API to serve data, if you have the resources to do so.
	24. Use Web Standards as the foundation of APIs	When designing APIs, use an architectural style that is founded on the technologies of the web itself.
	25. Provide complete documentation for your API	Provide complete information on the web about your API. Update documentation as you add features or make changes.
	26. Avoid Breaking Changes to Your API	Avoid changes to your API that break client code, and communicate any changes in your API to your developers when evolution happens.
Data preservation	27. Preserve identifiers	When removing data from the web, preserve the identifier and provide information about the archived resource.
	28. Assess dataset coverage	Assess the coverage of a dataset prior to its preservation.
Feedback	29. Gather feedback from data consumers	Provide a readily discoverable means for consumers to offer feedback.
	30. Make feedback available	Make consumer feedback about datasets and distributions publicly available.
Data enrichment	31. Enrich data by generating new data	Enrich your data by generating new data when doing so will enhance its value.
	32. Provide Complementary Presentations	Enrich data by presenting it in complementary, immediately informative ways, such as visualizations, tables, web applications, or summaries.

(continued)

Table 4.3 (continued)

	Best practice (World Wide Web Consortium, 2017)	Description (World Wide Web Consortium, 2017)
Republication	33. Provide Feedback to the Original Publisher	Let the original publisher know when you are reusing their data. If you find an error or have suggestions or compliments, let them know.
	34. Follow Licensing Terms	Find and follow the licensing requirements from the original publisher of the dataset.
	35. Cite the Original Publication	Acknowledge the source of your data in metadata. If you provide a user interface, include the citation visibly in the interface.

World Wide Web Consortium (2017)

4.5 Conclusions

In sum, opening government data is not easy, and there are many aspects that need to be considered when a public agency decides to open datasets. In this chapter we identified 11 organizational issues for opening up government data. These encompass six data-related issues (potential privacy breaches, data sensitivity and security, embargo period, data openness, lack of control over its usage and lack of trust in the data user, data quality, and data documentation) and five infrastructure and process-related issues (lacking infrastructure and resources, unclear or shared ownership, changes to organizational processes required, negative consequences for the government, and benefits obtained by others than the government).

When governments consider opening their data, they need to make a trade-off between the potential benefits and the potential disadvantages of this decision. A key question is: to open or not to open the data? The data requires a trade-off in which either the benefits or risks of opening may dominate. Figure 4.1 shows the decision-making process in which the benefits and disadvantages of opening data are weighed. Some data has many benefits and hardly any disadvantages and can be opened without any discussion. Other data should not be opened without any doubt due to security, privacy, or other reasons. There is a huge pile of data requiring a trade-off in which either the benefits or risks may dominate.

We do not know how large this part is that organizations need to decide on. Furthermore, it is likely that this changes over time. Since public values represent the needs and preferences of the collective citizenry, public values may change over time, as the needs and preferences of citizens may also change. It is likely that the decision regarding which data should be opened or closed will vary over time.

Thus, the most important trade-off is to open or not to open the data. This trade-off is based on the considerations that we described, such as data quality and data sensitivity. For each of the considerations, the civil servant responsible for data release needs to decide which aspects are more important. For instance, is it more

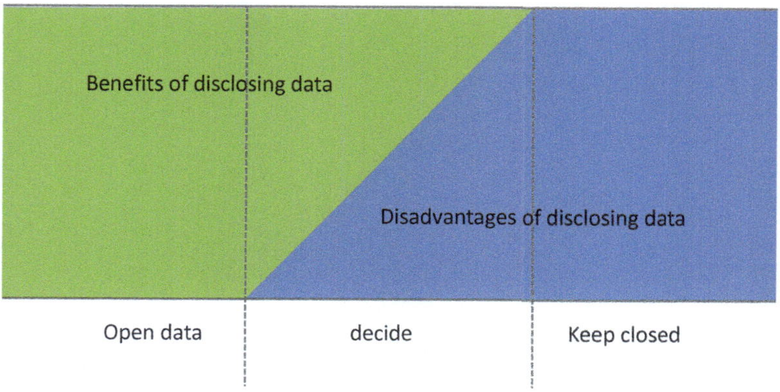

Fig. 4.1 Decision-making to open or not to open datasets (Zuiderwijk & Janssen, 2015, p. 114)

important that data are of high quality or is it more important just to publish the data and to let data users point out aspects of low quality? Is it more important to ensure that absolutely no datasets are published which are sensitive, and to remove all potentially sensitive variables? Or is it more important that the data is more useful, but might potentially be sensitive when combined with other data?

This chapter also provided several use-cases that describe how some of the identified issues can be overcome. The use-cases focused on solutions to reduce the risk of privacy violation (from an administration view) and on solutions to develop an open data infrastructure that enhances the coordination between open data actors (from a research view). Furthermore, we examined best practices as provided by the PSI-Share project and by the World Wide Web Consortium. Following these best practices should make it easier to reap the benefits of open data, as described in Chap. 1 of this book.

Chapter 5
Open Data Interoperability

"Semantic technologies enable open data interoperability beyond the point of pure format and structure alignment."

5.1 Interoperability in a Highly-Dynamic Open Data Ecosystem

The rapid growth of information technology during the last decade has put governments and businesses alike in front of a number of barriers to overcome in order to tap the full potential of this new digital era. One of the most challenging, but also most potential developments, comes with the web of data (Auer et al., 2007) and the inherent mass of freely-available information, i.e., open data (Zeleti, Ojo, & Curry, 2016). Especially open government data (OGD) holds the power to unlock innovation in both sectors, government and business, regarding the development of new, better, and more cost-effective services for citizens (Zuiderwijk & Janssen, 2014a). This interaction of actors forms a highly-dynamic ecosystem of data (Hammell et al., 2012), yet has to be re-evaluated with the increasing voluntary contribution of data by citizens, e.g., through citizen science initiatives (Lampoltshammer & Scholz, 2016) and open science data initiatives in general (Karmanovskiy, Mouromtsev, Navrotskiy, Pavlov, & Radchenko, 2016). Thus, approaching this ecosystem of open data from a quadruple helix (Carayannis & Rakhmatullin, 2014) approach is the next logical step. Figure 5.1 shows such an extended version of the ecosystem.

1. **Open Government Data** – this refers to data that was collected or produced within the public administration and the public sector in general. However, data affected by legislation, such as data privacy or national security, are not included.
2. **Open Business Data** – this refers to data that was collected or produced within the private sector, e.g., by organizations or companies. Its degree of openness

Y. Charalabidis et al., *The World of Open Data*, Public Administration and Information Technology 28, https://doi.org/10.1007/978-3-319-90850-2_5

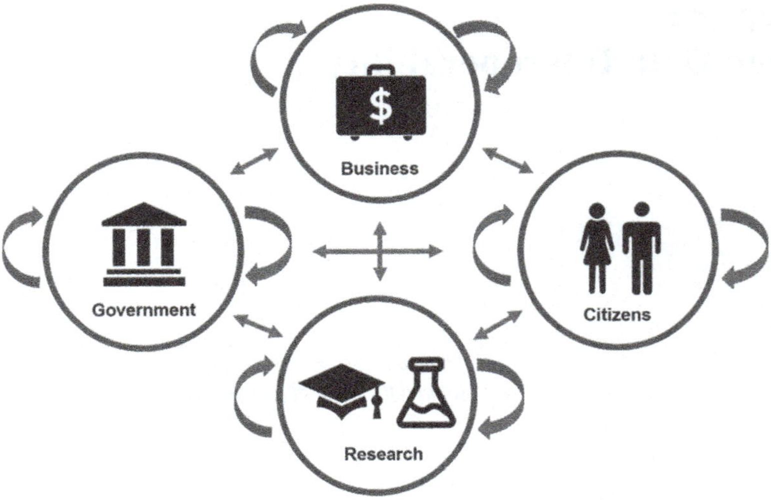

Fig. 5.1 Quadruple Helix-based open data ecosystem

and availability strongly depends on contract-based or sector-specific restrictions, put on the data by their producers.

3. **Open Citizen Data** – this refers to data regarding personal and non-personal related data of individual citizens. Examples can be found in the area of social media platforms or citizen science projects.
4. **Open Research Data** – this refers to data, which was collected or produced within academia and research sectors. It includes, e.g., publications or raw research data originating from interviews or experiments.

Obviously, this ecosystem introduces a certain level of complexity regarding the exchange and therefore also the interoperability of open data from the involved stakeholders. When discussing interoperability of open data, several levels can be distinguished in order to approach this issue via a technology-oriented, holistic way. According to Janssen, Estevez, and Janowski (2014b), the following four main levels of interoperability can be defined:

1. **Technical** – this level refers to a network-based interconnectivity between systems in order to be able to exchange data, e.g., on a per-transaction basis or via real-time streaming. By employing X-as-a-Service (XaaS) approaches, incompatibilities such as different operating systems or programming languages can be resolved.
2. **Syntactic** – this level refers to the use of standards in terms of exchange formats, e.g., XML or JSON, on a web interface level, i.e., for web services to exchange data.
3. **Semantic** – this level refers to reducing ambiguity in terms of data interpretability. This in turn requires semantic technologies and well-defined metadata, e.g., via ontologies.

4. **Pragmatic** – this level refers to quality and trust from an overall organizational perspective, including, e.g., service level agreements (SLAs) or context sensitivity in terms of meaning and involved stakeholders.

While all four levels are important to achieve a holistic approach towards the interoperability of open data, this chapter focusses on two of these levels, the semantic level and the pragmatic level, i.e., linking data as well as metadata and data quality.

5.1.1 A Semantic View on Data Interoperability

The World-Wide-Web (WWW) literally contains billions of pieces of information, spread out over a plethora of websites and information silos. This situation becomes challenging, when we are considering the search and retrieval of particular pieces of information. Thus, this unstructured way of storing information, e.g. as HTML pages, will – on the long run – not be sustainable. To counter this issue, the Linked Data paradigm arose, striving to interlink data on the web, pushing a new way of data handling towards the establishment of a semantically-enabled version of the WWW.

A way of describing this new version of WWW was originally provided by Berners-Lee via his Semantic Web Stack. The stack has become to some degree a blueprint for numerous implementations along the principles of the Semantic Web. Yet, the stack also visualizes the web from a high-level point of view, leaving open some important aspects and technology-related challenges yet to overcome. It is therefore no surprising that the stack has undergone several changes since it was first proposed. Figure 5.2 depicts a contemporary, but not necessarily comprehensive and final version of the stack. To provide a better understanding of the semantic stack, the following part introduces and describes the core layers, together with the core components of the stack (Hogan, 2013):

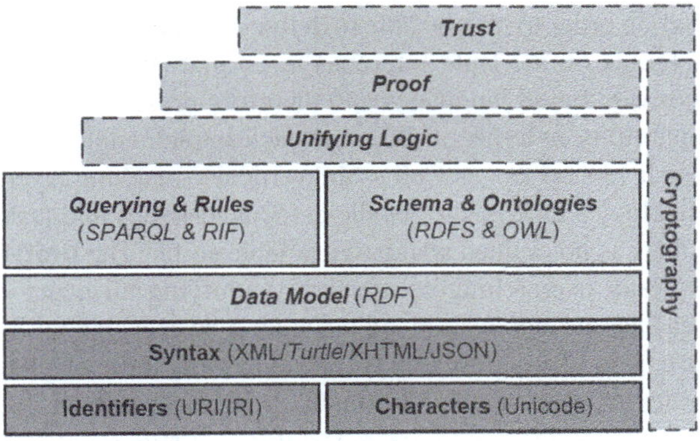

Fig. 5.2 Semantic Web Stack (Hogan, 2013)

The fundament of the stack is comprised of two elements. The first element is represented by mapping streams of data and external storage to actual textual information via the utilization of **characters** out of the Unicode char-set. The second element presents the ability to provide unique **identifiers**, which is imperative, considering the requirement for search, retrieval, and interlinking of resources in a machine-comprehensible manner. For the realization of provision of identifiers, the original stack foresaw the application of the Uniform Resource Identifier (URI), while current implementations shift towards a more general and flexible representation via the Internationalized Resource Identifier (IRI), based on Unicode. The next layer focusses on **syntactical** aspects, in particular, the provision of automatically parse-able elements, i.e., a common syntax in form of XML and JSON. While these classical forms are widely-adopted, custom syntaxes via, e.g. the TURTLE syntax (associated to RDF), are also possible.

On top of the syntax layer resides the **data model**. To provide the necessary means of data exchange, a common and machine-readable data model must be defined. This data model needs to be generic in that sense that it allows for the adoption of any content, originating from any given domain, while at the same time it must be usable without the need of proprietary technology. During the design of the Semantic Web, the Resource Description Framework (RDF) (Pan, 2009) has been chosen to serve as core data model.

Within the next layer, two components reside, which are required to introduce semantics into the Semantic Web. As RDF is only handling the structure of the content, but adds no semantic description to it, a formal way of additive modification to the existing model must be provided. This modification comes in form of formal languages, including meta vocabulary. The two basic variants contained within the stack are either the RDF Schema (RDFS) (McBride, 2004) or the Web Ontology Language (OWL) (Horrocks, Patel-Schneider, & Van Harmelen, 2003).

As it is the entire purpose of Linked Data to increase access and availability of data, there must be a way to search for these data by formulating **queries**, filters, and to design and apply search patterns in order to be able to identify data, as well as associated data, of interest. To realize this functionality, complementary to RDF, the SPARQL Protocol and RDF Query Language (SPARQL) (Quilitz & Leser, 2008) developed. In order to also be able to define certain sets of rules, the Semantic Web currently builds on the Rule Interchange Format (RIF) (Kifer, 2008), which covers numerous rule-based languages and therefore provides a high level of flexibility and compatibility in terms of different stack implementations.

For following layers on top, as well as the vertically-reaching layer, an increasing amount of technologies emerges to handle associated issues and tasks within these elements. Yet, there is no defined standard available so far. The **unifying logic** layer strives to provide an overarching compatibility, unifying all query languages and knowledgebases via the application of a comprehensive and unifying language. While there have been several research works addressing these challenges (Gyawali, Shimorina, Gardent, Cruz-Lara, & Mahfoudh, 2017; Krötzsch, Maier, Krisnadhi, & Hitzler, 2011; Polleres, 2007; Straccia & Bobillo, 2017) none of them was able to achieve a "one size fits all" solution up till now. The concept of a layer of **proof** is dedicated to the idea that the combination of various and externally-hosted data sets is a complex process and therefore has to provide some way of re-assurance for potential users of a stack implementation. This also holds true regarding applied reasoning processes, filters, or task completion. The **trust** layer is directly-connected to the layer of proof. Potential users or machine clients should be able to evaluate, if and to what degree they are able to trust certain agents providing data as well as resources and results, based on issued queries. Classical approaches use white-listing or black-listing, which in turn triggers the question, who is going to be responsible for maintaining these lists and therefore keeping them up-to-date. This again would push the issue of a central authority, which to some degree might compromise the entire idea of a distributed resource network. Finally, the **cryptography** layer is envisioned to integrate security and controlled access as cross-cutting concern throughout the entire stack. Aspects to be covered by this layer include the possibility to establish encrypted connections via secure protocols or the application of crypto algorithms such as RSA or AES to guaranty protection and privacy of data, information, the requests and search queries respectively. Furthermore, the layer also provides means of controlling, who can find, query, and finally access linked resources.

5.1.2 A Schema View on Data Interoperability

Besides approaching the topic of data interoperability from a semantic point of view, one can also refer to it through an architectural point of view, expressed by metadata schemata. Zuiderwijk, Jeffery, and Janssen (2012a) suggested the following three layer-based metadata architecture approach, as shown in Fig. 5.3.

The first layer enables to initiate queries for Linked Open Data, while the second layer provides enriched information regarding the dataset of interest, such as

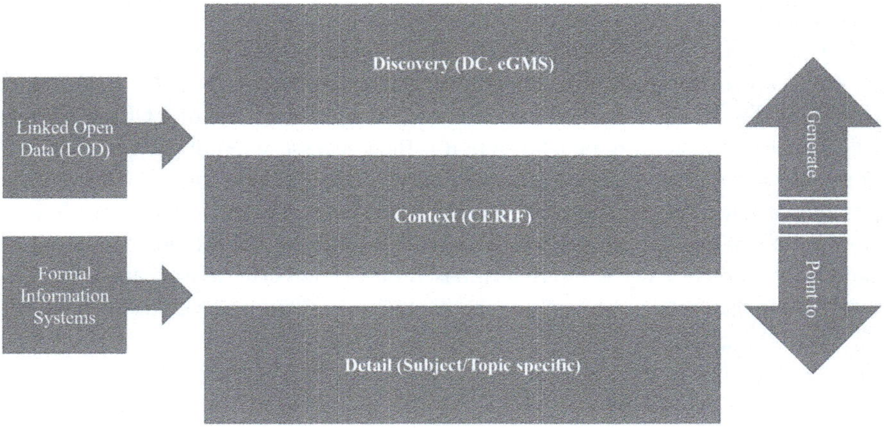

Fig. 5.3 Three layer-based metadata architecture. (Adapted from Zuiderwijk et al. (2012a))

involved persons, organizations, publications etc. At the same times, this layer is also responsible for the identification and generation of common metadata information to achieve a high-level of congruence. The third layer features metadata information which is specific to a domain, such as the Infrastructure for Spatial Information in the European Community (INSPIRE) (Directive, 2007). Within the first layer, several types of metadata standard descriptions can be applied, such as Dublin Core (DC)[1], the e-Government Metadata Standard (e-GMS)[2], or the Comprehensive Knowledge Archive Network (CKAN)[3]. The level of reduced complexity in these standards allow for an eased mapping process. Yet, this comes at a cost, namely, the used vocabulary not meeting necessarily the real-world demands, and compromises have to be made, which could after all results in poor query results or datasets not being discovered at all. It is due to this reason, why the second layer incorporates a layer of contextual metadata, expressed by the use of CERIF[4]. By doing so, the establishment of relationships between entities becomes possible. In addition, CERIF is the recommended metadata standard by the EC to be used by its Member States. Finally, the third layer allows for the attachment of highly-specific metadata, e.g., information about the domain, in-depth descriptions of the actual data, about the data collection process, etc. It is due to their important task of providing interoperability that metadata schemata play a significant role within the process of setting up a data infrastructure. For more information regarding data infrastructures, please refer to Chap. 6.

5.2 The Data Life-Cycle Within the Semantic Web

According to Auer, Lehmann, Ngomo, and Zaveri (2013), the following steps are required to form a complete data life-cycle (see Fig. 5.4) in the domain of Linked Data. It has to be noted though that while the cycle forms a kind of sequential order of steps, these steps may also occur in different combinations, depending on the current status of the resources under observations.

To begin with, any unstructured representation in form of, e.g., data sets have to be transformed in order to be compatible and map-able via the RDF data model (EXTRACTION). This process continuous until a critical mass of RDF-based data has been accumulated. In the next step, it is then necessary to not only provide sufficient storage for the collected data, but to provide features such as indexing and the possibility to formulated and apply search queries on to the data as well (STORAGE & QUERY). While current systems are already capable of interlinking data semi- or even fully automatically (LINKING), based on defined criteria and attributed features within data sets, it is essential that manual link creation as well as the possibil-

[1] http://www.dublincore.org/

[2] http://www.agls.gov.au/links/

[3] https://ckan.org/

[4] https://www.eurocris.org/cerif/main-features-cerif

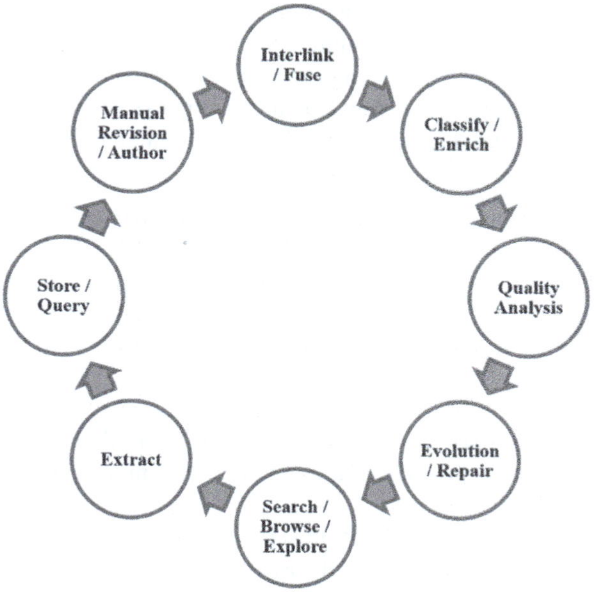

Fig. 5.4 Linked data life-cycle. (Adapted from Auer (2011))

ity to modify existing links is provided to further improve and refine the growing network between the data resources (AUTHORING). Yet, linking existing data sets and resources is not enough. Theses established links are per se not revealing any additional information regarding the classification of data sets or resources, nor are they providing knowledge about inherent structure as well as associated schemata. Therefore, the enrichment of data with high-level information and semantics is imperative (ENRICHMENT), to be able to increase the level of efficiency regarding aggregation and, in turn, towards searching and querying the growing semantic network. While identification and retrievability of data sets and resources is important, the results as such do not provide any information regarding the actual quality of the data or the associated metadata. Therefore, functionalities and services must be established to analyze the linked data and to identify potential errors or missing pieces of information within these data sets. Hitherto, for the services to work effectively, they require a well-defined set of quality metrics, describing what the term data quality implies for the given type of data (QUALITY ANALYSIS) – a detailed overview of such metrics can be found in Chap. 8. Once open issues are identified, smart algorithms can then be applied to correct these errors or, in some cases, even to reconstruct missing data pieces and therefore information (EVOLUTION & REPAIR). The last step then covers the usability of the entire system and Linked Data network by potential users (SEARCH, BROWSING & EXPLORATION). The best and most refined data corpus is of no use, if users are not able to efficiently browse through the data structure, intuitively formulate questions in form of queries and patterns, as well as to retrieve the desired information. Furthermore, smart systems will not only detect results that match user queries 1:1, but also allow for a certain form of fuzzy queries, providing users with potentially interesting alternative search paths and therefore leveraging the full potential of Linked Data.

As the presented cycle is of iterative nature, it is per se never completed and thus continuously leads to the improvement of Linked Data and in the long run, offers several benefits such as (Auer et al., 2013):

- *Uniformity*: as all data sets have undergone the transformation process from non/semi-structured data towards structure data into the RDF data model, the benefits of the RDF structure can be exploited. As all facts within this data model are formulated as triples formed by subjects, predicates, and objects, theses directly correspond to the applied unique identifiers (i.e., URI/IRI) and therefore reduce ambiguity.
- *De*-referenceability: via the application of the afore-mentioned unique identifiers, entities within data sets cannot only be precisely defined, but at the same time, serve as links between resources on the web, similar to URLs used to navigate between HTTP resources.
- *Coherence*: the core data model RDF supports the use of so-called namespaces. These namespaces allow for multiple use of identifiers without causing conflicts in terms of ambiguity. For example, the subject-predicate-object structure allows the establishment of links of entities between different namespaces via their URIs.
- *Integrability*: as the RDF data model provides uniformity across all transformed data sets, it becomes possible to build upon this unified structure to attach additional schema information or semantics in terms of ontologies. By doing so, the level of expressiveness of queries and answers can be significantly increased, which in turn enables and improves a more sophisticated matching process.
- *Timeliness*: the underlying process of publishing Linked Data is, due to the existing tools and technologies, relatively straightforward. In addition, once a linked data set has been updated, the process of accessing the newly-added information is easier, compared with the alternative way involving complex procedures in course of ETL (extract, transform, load) task.

An in-depth discussion regarding the single steps of the cycle, including the required tools and methods can be found in Chap. 2, paired with a comprehensive overview of different use-cases of the data life-cycle.

5.3 Ontologies as Means of Providing Semantics

The term "ontology" takes different meanings throughout different disciplines. Approaching the origin of this term from a philosophical point of view – the "big O" ontology – it can be described as a set of types and associated structures of objects, combined with properties, processes, all in relation towards every aspect of reality (Smith, 2003). Within the domain of computer science, one of the most referenced definition is provided by Gruber (1995), who sees ontologies as a formal way to explicitly specify a conceptualization and share it with others as a simplified representation of the real world for a specific purpose.

Ontologies haven been applied in a variety of application domains, such as the automated generation of user interfaces based on Linked Data (Hitz, Kessel, & Pfisterer, 2017), the detection of discriminatory language (Salguero & Espinilla, 2018), the classification of objects in satellite imagery (Lampoltshammer & Wiegand, 2015), the implementation of content management systems in the field of curricula development (Olteanu, Ionita, & Solomon, 2017), for the purpose of requirements engineering (Dermeval et al., 2016), as well as for data management in general (Daraio et al., 2016). Yet, this plethora of potential application domains also comes along with some drawbacks. Firstly, one of the most significant issues during the design and development of ontologies can be found in the so-called "semantic gap" (Smeulders, Worring, Santini, Gupta, & Jain, 2000). This term describes the difficult situation of providing detailed and concise description of visual interpretations. Although this example is strongly-related to the image interpretation domain, it well exemplifies the challenge of formalizing an objective view on reality, which is discussed in philosophy since decades, also known as the paradigm of "constructivism" (Jonassen, 1991). Besides this hurdle, ontology design and development suffer from the same issue, already known from knowledge modelling, such as overfitting (Hawkins, 2004). Overfitting occurs, if the knowledge model includes more features than necessary to describe a certain concept properly. This situation can arise, if the data set, which is used for the modelling, contains attributes and features, which are not representative for the kind of data at hand, but are present, e.g., due to errors within the actual data.

Yet, not only the process of designing and modelling of ontologies is a challenging task, the process of integrating and joining ontologies on different levels within one domain, or across domains, generates pitfalls as well. In addition to the beforementioned challenges, the following problems have also to be considered (Zhao & Ichise, 2014):

- **Ontology heterogeneity problem:** As data sets are published within a Linked Data environment, one part of the publishing process is to interlink these newly published data sets with already existing data sets. Yet, there is no existing "jack of all trades" ontology, meaning that the controlled vocabulary is nowhere close to completely cover all aspects of the interlinked data sets at once. Amongst other dimensions, two particular aspects increase the level of difficulty during the integration process. The first aspect addresses terminological issues. For example, one particular entity is modelled and described differently between ontologies foreseen to be integrated ("startingDate" vs. "beginningDate"). The second aspect focusses on conceptual issues, namely, entities differ in their hierarchical position within the ontology, as they were modelled in each of the ontologies as children of different parents, and therefore originate from different core concepts.

- **Identification of core ontology entities:** real-world entities based on the class descriptions including their attributes and properties within an ontology are called individuals. If the ontology and the included instances are of high volume, the identification of essential core properties of a specific class becomes increasingly difficult. To tackle this issue, the observation and notation of commonly-used core classes can support developers in their task to describe instances of particular data resources. Via these core entities and their associated attributes and properties, it becomes possible to design and construct suitable SPARQL queries, closing the gap regarding missing pieces of information within data sets.
- **Missing domain or range information:** the underlying relation between classes and properties within an ontology is expressed via domain information in the RDF core data model. This information describes the suitability of properties to be used for instances of certain classes. In addition, range information, also within the RDF core data model, help to better comprehend data sets in terms of the included values. Yet, in a real-world environment, ontologies are often missing this crucial information regarding domain and range, which in turn renders the process of integrating different ontologies based on their classes and properties more difficult.

The research community currently works towards potential solutions to the aforementioned challenges. For example, Lampoltshammer and Heistracher (2014) proposed a workflow for classification of data instances with use of a dedicated plugin for the ontology modelling environment Protégé (Gennari et al., 2003), called OWLET. This plugin enables ontology modelers to import external data as instances into their ontology model for classification of these data items. Furthermore, the suggest approach can also be used for testing newly design ontologies, by using gold-standard test data and evaluating the classification results as well as the level of coverage regarding the included classes as well as associated properties. In addition, this evaluation approach enables designers to also verify the existing range and domain information, which is an essential step towards lowering the bar of integrating other existing domain ontologies.

Another research work comes in form of the Framework for InTegrating Ontologies (FITON) by Zhao and Ichise (2014). It also addresses the heterogeneity issue, as well as the difficult task regarding the identification of core entities as well as to provide the crucial information considering domain and range for ontology properties. The authors achieve this via the combination of three approaches (see Fig. 5.5):

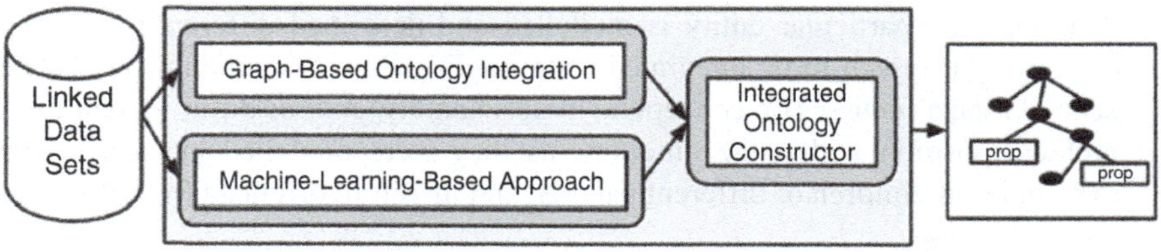

Fig. 5.5 Core components of FITON. (Adapted from Zhao and Ichise (2014))

- Step 1 – Ontology Similarity Matching on the SameAs Graph Pattern: during the process of integrating ontologies, 2:n ontologies are merged to deliver one unified model. Yet, in cases of small numbers of links regarding classes or properties, alignment becomes a challenging task. The authors therefore apply a WordNet-based (Pedersen, Patwardhan, & Michelizzi, 2004) approach, to establish undirected graphs between linked instances, which in turn provides valuable information regarding forming patterns between concepts over different data resources. These patterns can then be used to identify matching concepts to foster and speed-up the overall integration process.
- Step 2 – Machine Learning for Core Ontology Entity Extraction: to identify core entities within a given ontology, the authors apply machine learning algorithms. These algorithms comprise different approaches, starting out from rule-based classification via a priori knowledge, up to learning entirely new rules based on a data-driven approach.
- Step 3 – Automatic Ontology Enrichment: to be able to comprehend and understand the relationships between entities in the ontologies of observation, the domain and range information has to be seen crucial. Consequently, it is the next logical step to include this information during the integration process. The authors therefore take random samples out of the entire set of instances within the ontology and analyze their range and domain information via inspecting the associated properties and values. These results, paired with available standard range and domain information, is then used for annotating the resulting integrated ontology.

Considering the before-discussed complexity and depth of creating and maintaining linked data sets, the results will only be as good as the quality of the provided (meta) data, used to construct the actual links between the data sets. If the overall (meta) data quality is poor, linking of data sets may be not possible or might end up in erroneous links. Therefore, the next section will discuss the importance of quality aspects of Open Data and means to assess and evaluate quality of (meta) data.

5.4 Quality Aspects of Open Data

The overall quality of data sets is of upmost importance for several reasons. One reason is that without proper meta data and data quality, it is hard for experts to design and construct suitable ontologies for the domain the data set belongs to, due to missing information. Furthermore, this missing information, paired with potential errors within the data and the meta description itself can lead to false classification and therefore false linking or even no linking at all, as no common denominator as basis for the linking process could be identified.

The study conducted by Vetrò et al. (2016) identified several generic issues that can negatively affect the quality of Open Data (see Table 5.1). The first

Table 5.1 Potential data quality issues in open data sets

Incomplete data	Format not compliant to well-known standards	Lack of data source traceability
Incongruent data	Out-of-date data	Lack of metadata
Errors	High time to understand data	Lack of modification traceability

Adapted Vetrò et al. (2016)

issue is related to the data being **incomplete**. This leads to the metadata not matching, e.g., the time range of the actual data, which in return would deliver no matching data to search results of users. In addition, with the data being incomplete, analyses on this data is prone to produce wrong or misleading results.

The second issue comes in form of the actual data format **not being compliant** to well-known standards. This can cause problems from several directions. On the one hand side, automated data extraction, transformation, and loading (ETL) processes become difficult, if not impossible, due to the data not adhering to known and well-define structures and schemata. On the other side, the data as such might require special software to work and to incorporate them into existing data infrastructure and therefore acts as impediment for adopting the data. This manifests itself through additional costs for users as well as potential issues for long-time preservation of data, as proprietary software might not be available in the future. The third issue is present through the **lack of traceability** regarding the origin of the data at hand. This is not only a problem regarding potential licensing issues, but also in terms of contacting the original author(s) of the data, in case errors or gaps in the data have been identified and could be reported back to fix these. The next issue comes along in terms of **incongruent data**. This problem usually arises when data is merged, and the particular data set was not aligned to use the same format or schema. Thus, data items can have mixed data representations such as different date formats (Linux timestamp vs. date-time format). In consequence, filtering and/or sorting of data, as well as providing statistics regarding the actual content of data set becomes burdensome and only possible, after an additional step of type conversion. Next issue on the list is present by the data being **out-of-date**. An example would be a data set containing scheduling information regarding a certain type of public transportation, e.g., bus lines. Such public transportation information often changes slightly from 1 year to the next, thus, if the data set called “bus schedule Vienna” is not updated accordingly, this leads to issues regarding the use of this data in, for instance, customer apps for public transportation. Further issues are present in the lack of metadata. In cases, where **no metadata** is available at all, mapping and interconnecting of data becomes only possible, after going through the data themselves, which can be a time-consuming and costly operation. Also, an assessment regarding schema or format compliance, as well as the application of other metrics is not straightforward, same goes for indexation of datasets. Another common issue is found in **errors** directly within the data themselves, or within the associated metadata. Of course, if the data at hand are incorrect, analyses of these data will produce erroneous results as well. An often neglected but still important issue comes with a **high time to understand the data**. While the data themselves can be complex, the understanding of them can be eased via meaningful descriptions and annotations by a complete set of metadata. If this description is missing, it is sometimes not even possible to determine, what the data is about, what is their range, and what details are included in the data set at hand. Finally, there is the issue that comes along with a **lack of modification traceability**. While the origin of the data as well as their producer can probably be determined via the associated metadata, changes within the data are not obvious. If not provided with a set of history or changelog, detecting modifications, additions, or removal of a single datum or even complete sequences of data are impossible. Thus, manipulation or unintended data loss cannot be detected or proven.

As all of these issues can fairly impact the usability and adoptability of open data, numerous research projects are focusing on assessing the quality of open data via the introduction of metrics as well as approaches to fix some of the identified issues automatically or at least provide support during the manual process of data cleaning and repair. Thus, the next section provides an overview of ongoing activities in that regard.

5.5 Quality Assessment and Improvement of Open Data

To identify suitable data sets for a particular application, their quality has to be assessed first. This assessment is usually performed via the use of so-called data quality dimensions and associated metrics (for an in-depth discussion see Chap. 8). According to Heinrich, Kaiser, and Klier (2007), well-defined metrics should match the following criteria:

1. **Measurability** – being defined quantitatively, normalized, at least interval-scaled
2. **Interpretability** – specific focus to increase comprehensibility
3. **Aggregation** – quantification on attribute level, while keeping semantic consistency across all levels, to enable cross-level aggregation
4. **Feasibility** – clearly defined input parameters, while at the same time providing a high level of automation

Alongside these basic preconditions, researchers have developed various approaches regarding the assessment of data quality. The work by Borovina Josko and Ferreira (2017) presents a case study regarding the use of visualization approaches to enable data quality assessment to identify defects in the structure of the observed data. Debattista, Auer, and Lange (2016) introduced the Luzzu framework as a generic approach to assess the quality of linked open data. Luzzu consists out of four main components, namely a flexible interface to enrich the framework with new assessment metrics if required, an ontology-driven backend regarding metadata quality representation, a scale-able stream processor as end-point for, e.g., SPARQL endpoints, and a user-defined ranking algorithm. Kontokostas et al. (2014a) adopted the idea of test-driven evaluations out of the software engineering domain into the task of assessing the quality of Linked Open Data. The authors leverage a large collection of test patterns, derived from SPARQL queries to conduct their test runs. Acosta et al. (2018) applied an innovative solution towards the quality assessment of Linked Data via a crowdsourcing approach. Crowdsourcing in this case means that a large group/network of people, which are not pre-defined, are working towards a common task or goal. Crowdsourcing has established itself in many different areas, starting from microtask working (e.g., Amazon Mechanical Turk), to funding projects of common interests (e.g., Kickstarter). Usually, the tasks put towards the crowd are single-iteration based, yet there are also approaches building on multiple iteration to assess and evaluate the results from the crowd by the crowd itself. Acosta et al. describe three main ways of crowdsourcing on a given task:

Contest-based Crowdsourcing follows the idea of handing over a particular task or problem to solve to the crowd and in consequence to reward the best, most efficient, most effective, or most innovative solution (Leimeister, Huber, Bretschneider, & Krcmar, 2009). The approach leverages on the exploitation of intrinsic motivational factors, triggered by competition and intellectual stimuli. The contests are usually held open for an extended period of time – depending on the complexity of the task – to allow for enough time to submit a solution to the described problem. While there are several ways of stating a reward to the best solution, usually a main prize is provided by the entity that issues the challenge. While these challenges have been around for years to attract experts to work on a given problem, they are increasingly used towards working with citizens as well, and in consequence, also contribute towards the entire citizen science movement (Lampoltshammer & Scholz, 2016).

Microtask Crowdsourcing applies the approach of splitting a given problem into chunks, thus called microtasks (Howe, 2006). This approach works best if the abilities for solving these microtasks are either based on basic audio or visual comprehension, or towards the understanding and interpretation of language-related issues, rather than towards the necessity of a priori expertise in the related topic. In order to be handled in an efficient way, microtask crowdsourcing requires a high level of parallelization, and in consequence a large number of participants. Thus, this decentralized method results in faster responses, in conjunction with the possibility to validate the proposed solutions to the posed problem based on, e.g., majority voting or other consent-finding methodologies. Typical awards issued for successfully solving microtasks are provided in micropayments.

Crowdsourcing Pattern Find-Fix-Verify (Bernstein et al., 2015) similar to the microtask crowdsourcing splits a more complex task into a set of tasks of less complexity, which are then processed throughout three consecutive stages. In the first stake, the individuals within the crowd are to **find** data, which is of interest to solve the given task. In the following second stage, the outcomes of the first stage are corrected/amended (**fixed**) if required to match the given task in a better way. Then, in the third stage, the final results are verified one last time to conclude the overall quality assessment. This pattern does not only exploit the benefits of the before-described microtask, but also gains within each step of the negotiation process between all involved crowd members. Furthermore, alongside the three different stages, different compositions of crowds can be used to even more increase the likelihood of high quality output.

As discussed before, not only the linking of data supports interoperability, data quality does as well. Regarding the later, promising approaches have been found in regarding to the assessment of data quality via metrics as well as via leveraging the knowledge and the abilities of the crowd. From the given point of view, it is the next logical step to combine these two approaches to make use of advantage of both side in a synergistic way. The following section therefore presents two research projects and initiatives, which also build heavily upon the crowdsourcing aspect for the identification of data issues, paired with automated assessment and correction abilities for data quality and thus going towards the improvement of open data interoperability.

5.5.1 ADEQUATe Project

The ADEQUATe project was initiated to develop innovative approaches towards the measurement, monitoring, and improvement of date quality and to demonstrate these concepts via two pilot use-cases in Austria, i.e., data.gv.at and opendataportal.at (see Fig. 5.6). To achieve this ambitious goal, the project tackles the four main issues identified during its initial requirements elicitation phase (Höchtl & Lampoltshammer, 2016):

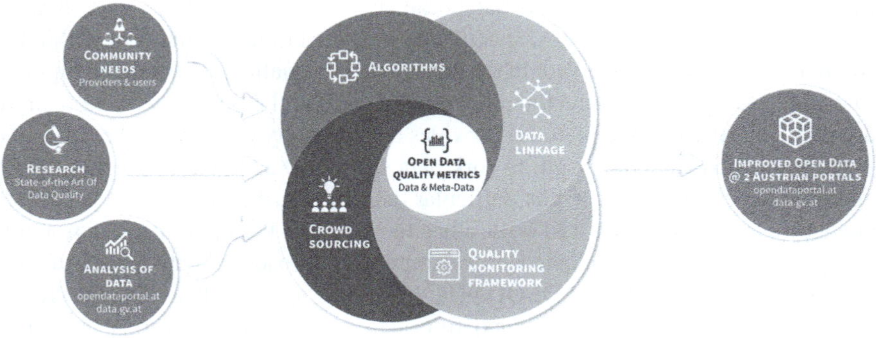

Fig. 5.6 The overall conceptual model of the ADEQUATe project. (https://www.adequate.at/)

1. **Issue – Defining suitable quality metrics targeted for open data**: as already discussed in the sections before, there do exist numerous metrics to assess data quality. Yet, often they do lack, besides still fulfilling the basic criteria of well-define metrics, the specific characteristics required by open data as well as the target platform and audience. Furthermore, applying all available metrics to a given data set may introduce an unjustified bias by falsifying the assessment results due to, e.g., important metadata fields missing, which results in a reduction of the overall quality score of the assessed dataset.
2. **Issue – Providing (semi-) automated improvement of metadata and data quality**: while identifying issues regarding metadata and the data as such is one aspect, the overall big picture would be incomplete without considering the automated correction of potential issues as well as further improvements towards the dataset and its associated metadata. Yet, this part is challenging in particular, as the algorithm itself has to decide what to change in order to improve the overall quality scoring. At the same time, improvements expressed by quality metrics do not necessarily reflect the possible introduction of content-wise errors by the system.
3. **Issue – Coping with CSV-based data sets**: one of the biggest challenges within the existing datasets of the two pilot portals are represented by data in the CSV format, as these data present the majority of datasets on the portals at this point in time. CSV files are known for their issues regarding proprietary formats, such as delimiters (depending of their source language. e.g., German vs. English), nested tables, or non-present metadata.
4. **Issue – Foster open data community engagement**: while algorithms may assess and correct potential errors within data, without the continuous feedback and expertise of the community, i.e., the end-users of the data, data providers, as well as service provider, building their services on top of the existing open data, no sustainable development can be realized.

To deal with these four main challenges, the ADEQUATe project combines community-driven solutions with state-of-the-art technologies in the domains of data quality assessment, correction, as well as monitoring. In a first step, the project continuously monitors the quality of open data being published at the two uses-cases, namely data.gv.at and opendataportal.at. This is achieved via a set of well-defined dimensions and metrics, specifically designed to match the data within the two data portals being observed. In the next step, data quality algorithms are applied to (semi)-automatically correct identified issues within the observed (meta)data. In addition, the ADEQUATe platform provides a community component, based on the well-established technology git, to fork data sets of interest and to resubmit fixed and/or enhanced versions of this particular data set. Furthermore, these suggested changes can then be discussed with other members of the open data community, making full use of the intended crowdsourcing approach. Finally, the semantic enrichment component of ADEQUATE, based on tools such as Odalic (Knap, 2017), tackles the open issue of existing legacy data and transforms them into Linked Data.

5.5.2 Openlaws

The linking of data provides increased access, transparency, and availability of information. This fact does not only hold true within the business and research domain, but also for public administration, which have an obligation and responsibility towards their citizens. In case of public administrations and governments, the distribution, availability and access towards legal information is imperative. Yet, there exist some severe issues at the moment regarding this access. One of them is found in form of available APIs, which are not always up and running on a 24/7 basis, paired with slow systems and often non-compliant data towards standard or even self-issued schemata. This in turn makes the use of automated crawling and analysis more than difficult. Translating this situation into a cross-border context, the problem becomes even bigger, as each member state within the European Union are providing their open legal data in different formats, often with metadata in their own language (e.g. the Netherlands) and not towards a better understanding in a common language such as English. To overcome these issues, the EU research project openlaws[5] and the resulting spin-off are built around three core pillars, namely open legal data, open source software, and open innovation towards the establishment of Open Justice in Europe through open access to legal information (Lampoltshammer, Guadamuz, Wass, & Heistracher, 2017). The project's main goal is to increase the level of access towards legal information by supporting users in organizing and sharing their respective information (Wass et al., 2013). Nowadays, a small number of organizations and companies sign responsible for publishing and distributing legal information. Yet, this distribution occurs in somewhat restrictive and non-transparent ways, e.g., through public governance bodies or through public-private-partnerships with certain established publishing houses. Due to this fact, the important access to metadata of legal data is also restricted, which hinders automated processing of these data. Within this often-commercialized ecosystem, legal experts publish their research work and knowledge, with little to none free information flow towards the public and wider research community. This stands in sharp contrast to other research areas, where open research data and knowledge is shared increasingly.

Openlaws tries to break this restricted circle and therefore supports citizens in accessing, working with, and finally understanding legal information and in consequence, their rights and responsibilities towards the state and society. But not only citizens can profit from the project's outcomes, companies and organizations do as well. Supporting them with the required information and knowledge regarding necessary legal compliance according to their field of business, the experts within these organizations and companies can contribute to the sustainability of their business model as well as demonstrate proficiency towards their customers and clients. In comparison to the existing environment, the newly established platform is all-inclusive, meaning that publishing house can as well offer and integrate their premium content, enriching the data at hand even more.

[5] https://openlaws.com/

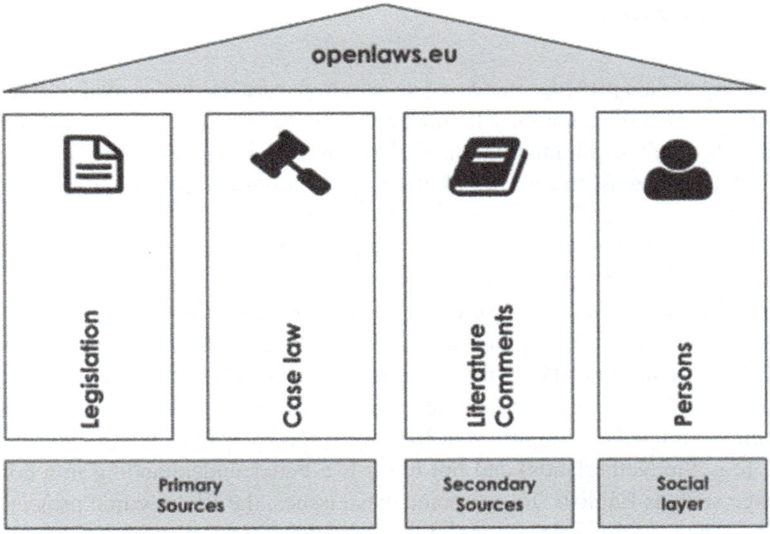

Fig. 5.7 Core components of the openlaws platform (Lampoltshammer et al., 2017)

Finally, public bodies and governments can push more than ever open legal information towards the community, following the idea and legal context of the public-sector information (PSI) directive.

To achieve this ambitious goal, the project provides the following services to its users, based on the core components shown in Fig. 5.7:

- The possibility to conduct a meta-search across several national legal databases and therefore provides cross-border and also cross-language access to legal information
- The amount of legal information is increased, providing additional possibilities for legal scholars and researchers to distribute their work, in direct context with the legal basis their working on and the audience there are targeting, who is affected respectively.
- An improvement of legal data and information quality, as experts can evaluate and curate the data within the platform, as well as the hosted publications in a new way of peer-review
- The existing network of legal scholars, experts, and practitioners is further extended and is also made available and searchable for citizens
- Finally, the access to, e.g., case law can provide a better understanding of laws, regulations and associated consequences for all affected stakeholders. Thus, the availability of open legal data and therefore the derived open legal information contributes towards better democracy and policy-making in the long run

To provide these services, the openlaws platform builds upon existing open data sources across the Union, such as national legal databases and EUR-Lex. These information are aggregated into Big Open Legal Database (BOLDbase), based upon an innovative graph database approach (Lampoltshammer, Sageder, & Heistracher, 2015). This new way of interlinking previously disconnected open legal data generates a new way of working with and providing legal information for all interested stakeholders. In addition, while experts and citizens are interacting on the platform with each other and with the legal data in openlaws, the platform makes full use of these interaction via integrated analytics, e.g., creating recommendations for individuals in regard to potentially-interesting legal information as well as additional benefits such as automated update services to broadcast important changes within legal domain of particular interest for each individual user.

5.6 Conclusion

Open data interoperability is imperative to drive the movement of Linked Open Data and therefore to increase not only the level of discovery and accessibility of data, but also the possibility to fuse data in order to create new application scenarios. These application scenarios can cover various stakeholders in a transdisciplinary way, including businesses, academia, public administrations, and citizens alike. Data interoperability is also the key for the exchange of data in different types of infrastructure (see Chap. 6), which can be seen as the key to enable the vision of the European Commission regarding the Digital Single Market. But interoperability is not only expressed by the application of common data formats and standards, the overall quality of the data itself and the associated metadata is also important, as these factors do not only impact processing of the data but usability of the data in general (see Chap. 8 for more about quality metrics and overall assessment). Overall, it can be stated that although the barriers of open data adoption have been known for a while, the "golden solution" is still missing to fulfil the high expectations that were expressed when the Public-Sector Information Directive (PSI) was put into place. Interdisciplinary research projects such as the openlaws project and the ADEQUATe project are an important step forward to increase accessibility of open data, focusing especially on data quality, as well as the semantic linkage of data to increase awareness on the one hand, but also adoption of available data on the other hand. The second aspect is crucial, if sustainable data-driven business models (see Chap. 7) shall push the European Union back into the international "game of data".

Chapter 6
Open Data Infrastructures

"User-centricity, transparency, and trust are the key elements towards a sustainable open data infrastructure."

6.1 Forming Open Data Infrastructure

Data represents a key asset in virtually any aspect of society and economy. Open Data in particular represents a source of immense value, as social capital (Lampoltshammer & Scholz, 2017) as well as an asset for business cases. Governments and their public administrations are generating and collecting during their service a plethora of different kinds of data, as well as an enormous amount in terms of volume. To tab into the potential this data holds in terms of stimulating economy, as well as the development and enhancement of governmental service for the benefit of the public (see Fig. 6.1), a sophisticate Open Data Infrastructure is required.

The Open Data Institute (ODI) sees data infrastructure as tangible and important as classical infrastructure, such as electricity or road networks. Data infrastructure have the main goal to keep the society informed and therefore contributes directly towards an increased accessibility and governance regarding data. Data within the infrastructure is quite heterogeneous, comprising not only governmental data, but also data from the business sector as well as data from non-profit organizations. The increased transparency in consequence can lead to not only business value, but also to environmental gains as well as towards societal benefits. In general, the ODI describes three different kinds of data infrastructure (Broad, Tennison, Starks, & Scott, 2015):

- **Local Data Infrastructure:** this kind of infrastructure contributes to an improved information state of citizens, communities, as well as decisions-makes on a governmental level
- **National Data Infrastructure:** this kind of infrastructure aims at strengthening the inherent resilience of a country in economic, social, and environmental areas.

Y. Charalabidis et al., *The World of Open Data*, Public Administration and Information Technology 28, https://doi.org/10.1007/978-3-319-90850-2_6

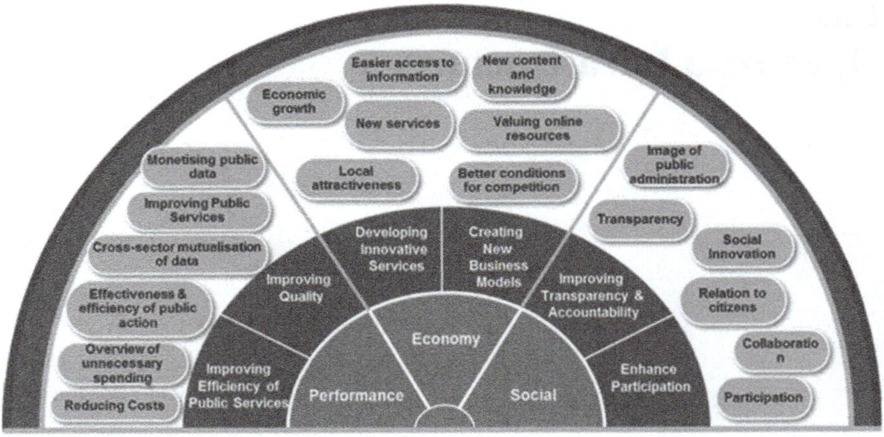

Fig. 6.1 Benefits of re-use of Open Government Data. (https://www.europeandataportal.eu/en/providing-data/goldbook/open-data-nutshell)

Besides the possibilities to build and provide services for citizens by companies and governments alike, the increased transparency boosts democracy as a whole.

- **Global Data Infrastructure:** this kind of infrastructure provides the means of tackling global issues such as getting insight to globally-acting entities such as multi-national organizations as well as a better understanding of progress regarding global policy-making.

With this important role of data infrastructure for individuals and society as a whole, there comes a great responsibility and requirements towards organizational and technological, as well as ethnical capabilities of organizations that provide these kinds of data infrastructure (Broad et al., 2015):

- **Long-term sustainability:** this kind of infrastructure contributes to an improved information state of citizens, communities, as well as decisions-makes on a governmental level,
- **Perceived authority:** citizens should hold a basic trust towards the maintainer of the data infrastructure, including its data,
- **Transparency:** the infrastructure should be transparent in a way that all processes regarding management and operations on the data themselves are well-documented and comprehensible, as well as replicable. Furthermore, the infrastructure should feature mechanisms, which allow for requests regarding an entities own data, what they were used for, who accessed them, etc.,
- **Openness:** the envisioned infrastructure should treat requests and users equally in terms of response, the right of information, as well as access to its inherent services and data, while at the same time protect the rights of individuals as required by law,
- **Commitment to the validity of data:** this attribution becomes most important in cases, where the infrastructure representing a de facto monopoly regarding

data storage and access to information. It should therefore be dedicated towards long-term sustainability of the data and information stored, paired with high aims regarding the provision of high-quality data, the use of standard formats, as well as its social responsibility towards the citizens,

- **Agility:** the infrastructure has to be able to not only handle the rapidly developing and demanding changes of technology and societal expectations, but also provide up-to-date data and information upon the request of external users.

All the before-mentioned criteria can be translated into a set of functional requirements, which should be fulfilled by an open data infrastructure. In the following section, we will have a closer look at these functional requirements, together with ongoing research activities regarding methodologies and tools to realize these requirements.

6.2 Functional Requirements of an Open Data Infrastructure

A sustainable open data infrastructure should reflect the needs and requirements of all involved stakeholders that are providing data to or using data from the data infrastructure. Zuiderwijk (2015b) conducted a research work towards the design of such an infrastructure to enhance the coordination of open data use. In particular, her study focused on the influential factor of OGD use, the functional requirements of an infrastructure for OGD, its functional elements, a concrete realization of such an infrastructure, and finally its overall effects. Table 6.1 provides an overview of the derived functional requirements of an open data infrastructure.

The requirements can be grouped within five main categories, namely, (i) searching and finding data, (ii) analysis of data, (iii) data visualization, (iv) interaction on this data, and (v) quality analysis of the data. In the following, we will have a look at current research works in these five respective categories.

Table 6.1 Functional requirements of an open data infrastructure

Category	Functional requirement
Searching and finding data	1. The OGD infrastructure should be a one-stop shop for datasets and metadata from a variety of other OGD infrastructures.
	2. The OGD infrastructure should allow OGD users to integrate and refer to datasets from various other OGD sources.
	3. Use controlled vocabularies to describe OGD.
	4. Use interoperable standards to describe OGD.
	5. The OGD infrastructure should support data search through keywords, data category browsing and data querying.
	6. The OGD infrastructure should support OGD use by the ability to search for data and metadata in multiple languages.
	7. The OGD infrastructure should facilitate filtering, sorting, structuring and ordering relevant search results.

(continued)

Table 6.1 (continued)

Category	Functional requirement
Analysis of data	8. The OGD infrastructure should provide data which describe the dataset.
	9. The OGD infrastructure should provide data about the context in which the dataset has been created.
	10. It should be clear for which purpose the data have been collected.
	11. It should provide examples of the context in which the data might be used.
	12. Domain knowledge about how to interpret and use the data should be provided.
	13. The OGD infrastructure should allow for the publication of datasets in different formats.
	14. The OGD infrastructure should offer tools that make it possible to analyses OGD.
	15. The OGD infrastructure should provide insight in the conditions for reusing the data.
Visualization of data	16. The OGD infrastructure should provide and integrate visualization tools.
	17. The OGD infrastructure should allow for visualizing data on maps.
Interaction on data	18. The OGD infrastructure should support interaction between OGD providers, policy makers and OGD users in OGD use processes.
	19. The OGD infrastructure should allow for conversations and discussions about released governmental data.
	20. The OGD infrastructure should allow for viewing who used a dataset and in which way.
	21. The OGD infrastructure should provide tools for interactive communications between OGD providers, policy makers, and OGD users (e.g. data request mechanisms and social media).
	22. The OGD infrastructure should provide tools for interactive communications between OGD users (e.g. discussion forums and social media).
	23. The OGD infrastructure should provide tools to keep track of amended datasets so that users know how datasets have been changed.
Quality analysis on data	24. The OGD infrastructure should provide insight in quality dimensions of OGD.
	25. It should be possible for OGD users, OGD providers and policy makers to discuss the quality of a dataset.
	26. The OGD infrastructure should provide information on the context in which a person reused a particular dataset.
	27. The OGD infrastructure should provide quality dimensions of datasets that are comparable with other datasets and with different versions of the same dataset.
	28. It should be possible to compare the quality of datasets over different data sources, over time and over data reuse on the data infrastructure.

Adapted from Zuiderwijk (2015a)

6.2.1 *Searching and Finding Data*

Sugimoto, Li, Nagamori, and Greenberg (2017) focused in their work on the topic of data archiving, especially metadata longevity. They provided suggestions and a proposed approach toward provenance of metadata registry in the area of risk management. In their work, the authors point out the challenges that arise from handling the context of the preserved metadata as well. This is a non-trivial problem, as the definition of concepts, which would be used to describe the context within a Linked Data environment, are prone to changes of time. Song (2017) proposed a method of linking data in the field of digital humanities across languages. This is achieved via use of metadata, yet without approaching the issue from the classical angle of translation. Instead, word embeddings are employed to then calculate a similarity metric based on the actual word vectors. The approach was successfully tested on a use-case involving Japanese and English. While there exists a plethora of shared vocabularies and ontologies, the actual engineering task of using them in a given context of a certain domain is challenging. Thus, precision regarding the description of concepts within an ontology is key. Out of this reason, Dutta, Toulet, Emonet, and Jonquet (2017) came up with a revised version of *Metadata vocabulary for Ontology Description and publication,* short MOD 1.2. This new version significantly increased the potential level of expressiveness of attribute-based ontology description, along with the possibility to semantic annotations via an OWL vocabulary to allow for the ontologies to be made available as Linked Data. When it comes to the task of creating Linked Data, e.g. in form of RDF, flexible and extensible tool are needed. To enhance current efforts in this research direction, Knap et al. (2018) introduced the UnifiedViews toolkit, an ELT framework that can handle a variety of associated processing tasks. Besides its capabilities of standard (pre-)processing tasks, custom modules can also be developed and integrated into the RDF creation workflow.

6.2.2 *Analysis and Visualisation of Data*

Kalampokis, Tambouris, and Tarabanis (2017) focused in their work on the combination of linked data approaches and open statistical data and the associated lifecycle. They created a toolkit named OpenCube, which allows for the associated actions specific to this data, covering its creation, expansion, and exploitation. Veith, Anjos, de Freitas, Lampoltshammer, and Geyer (2016) came up with a flexible, cloud-based solution for data processing and data fusion of heterogeneous sources, including open and closed data, based on a lambda architecture. By doing so, data of different temporal solutions and arrival speeds can be handled as well in various kinds of applications scenarios. The level of acceptance and adoption of open data strongly depends on the user experience delivered, while working and interacting with the data. An intuitive representation is key in this circumstance yet is hard to achieve due to the high level of heterogeneity of Open Data. Thus, Ojha, Jovanovic, and Giunchiglia (2015) introduced a methodology, comprising a novel visualization approach, based on the concept of treating data as entities. This goes along with preferences of users to group and sort items by exactly such entities. Paired with a tailored UI, the authors could successful demonstrate the increased level of user experience, while browsing and searching through Open Data catalogues. Speaking of data heterogeneity, this becomes also an issue regarding the process of data integration. This heterogeneity is found via various formats (txt, csv, pdf), as well as the inherent schemata or not existing schemata. The work of Carvalho, Hitzelberger, Otjacques, Bouali, and Venturini (2015) discussed the pitfalls along the way of integrating this data, especially in the realm of Open Data. The authors show ways of dealing with the arising issues, stressing and demonstrating the pivotal role of information visualization to guide and support users in the integration task. A unique approach towards the visualization of "human-sensed data" is proposed by McLean (2017). She collected data concerning smells and aromas reported by citizens, while walking through the city. Combined with the geographic location of these reports, a visual olfactory map was derived, for communicating the results to the public. This interesting approach towards data visualization offers insights to citizen-collected data and lowers the barrier of comprehension of information.

6.2.3 Interaction on Data

Interaction and feedback loops regarding the data itself as well as the use of the associated services of the infrastructure from the public are imperative for sustainable platform. Thus, it is necessary to understand, how online communities can be incorporated into innovative co-creation processes to further evolve the existing offering of data and services. Konsti-Laakso (2017) for example focussed her research on two main aspects, namely how these online communities can help in drafting and executing innovation processes within the public sector and second, what kind of role social media platforms take in this process, including the produced results. Also, in the context of Smart Cities, technology and Open Data play an important role in the development and successful growth of the urban environment. However, the pure existence of data is not enough. Gagliardi et al. (2017) stress in their work the necessity of the data being used, feedback gathered, and also distributed and communicated. To enable this communication loop between citizens and government, the authors developed, based on a design science research methodology, an ICT-based tool name *UrbanSense*. This tool is envisioned to foster the innovation process of new public services, by enabling information flow even on a real-time level between citizens and public administration. When dealing with the cooperation of public administration and citizens, democratic processes represent important impact factors. Ruijer, Grimmelikhuijsen, and Meijer

(2017) argue that existing open data platforms are over-simplifying these processes and therefore have failed so far to hold up to their promises. To overcome this issue, they developed a *Democratic Activity Model of Open Data Use,* covering monitorial, deliberative and participatory use-cases, advocating a context-sensitive design approach towards data transformation and interaction. A special focus on the interaction with the Open Data community is put by the Austrian research project *ADEQUATe* (Höchtl & Lampoltshammer, 2016). Here, the project realised a community platform that provides enhanced versions of open datasets from the two main open data portals in Austria. The community is not only informed about the overall quality of data, but can also jointly work on the improved datasets, discuss related issues and changes, as well as provide further improved versions back to the community. For further details about ADEQUATe, please refer to Chap. 5.

6.2.4 Quality Analysis on Data

The overall quality of data is not only important in terms of reusability, but also towards credibility when it comes to open governmental data. Torchiano, Vetro, and Iuliano (2017) developed a basic set of metrics to assess open governmental contractual data, based on the ISO SQuaRE standard in a way that the fulfilment and potential problems within the data can be identified automatically. Stróżyna et al. (2017) developed a framework for identification of suitable open data based on quality and availability aspects to be combined with internal closed data to increase the overall values for an organization or company. The authors see restrictions, e.g. regarding automated crawling, as one of the most dominant hurdles, besides the general quality of the available data. Thus, the term Open Data should in their point of view being revisited, as it does not apply to various resources available on the Internet. Mihindukulasooriya, García-Castro, Priyatna, Ruckhaus, and Saturno (2017) also address the problem of data quality, yet from the specific viewpoint of Linked data. The developed a RESTful web service called Loupe API that provides profiling capabilities for Linked data based on user-specified requirements. These requirements can cover explicit details such as RDF classes or vocabulary, as well as implicit requirements such as cardinalities between entities and multi-lingual aspects. The results can of their API can either be inspected manually or via dedicated validation languages such as SPIN. Further information regarding data quality metrics and assessment can be found in Chap. 8.

Besides all functionalities of a platform or data infrastructure, it will not reside without the trust of the users regarding the process being correct, the data hosted being valid, as well as their individual rights being protected. Thus, the next section puts its focus on the important aspect of trust and how modern technologies can enable trust in open data infrastructures.

6.3 Building Trust in Governmental Data Infrastructures

Trust in the governmental domain can be visited from two perspectives. The first perspective relates to the trust of citizens towards the public administration. If citizens trust the processes they are involved in, less feedback and personal interaction is required, which can result in reduced overhead and thus in less cost and time. The other perspective is the one of the public administration where monitoring and validating actions, documents, and information provided by citizens take time and produce costs as well (van de Walle, 2017). So, in order to approach trust form the viewpoint of both parties, a common technology-based approach to be incorporated into the data infrastructure has to be found. As one solution towards this issue, we will discuss the concept and applicability of blockchain technology.

6.3.1 Transparency Through Blockchain Technology

The overall concept of blockchain is basically a kind of database, which is hosted over a network infrastructure (e.g., Internet) in a de-centralised and distributed way (Ølnes, 2016). In particular, a blockchain is not only storing but on the same time updating all transactions that it stores over all connected nodes within the P2P network. On this network, all nodes can make use of it to store their transactions, with every party receiving its own copy of the transaction. It is noteworthy that nodes do not have to be actual human users but can be – along the paradigm of IoT – also machines and software services. Signing-up to this distributed ledger is possible via the use of public key algorithms. The validation is performed by all nodes, to build a consensus about the correctness of the submitted transaction. If a transaction is declared valid, it is stored within a block, which in return is added to the blockchain. Thus, the last added block also states the trust of the network towards the correctness of the current chain. Every block is a set of transactions including associated timestamps, as well as the hash of the previous block within the chain (Ølnes, Ubacht, & Janssen, 2017). A simplified summary of the main steps within a blockchain transactions can be seen in Fig. 6.2. Blockchain technology was also successfully introduced as secure information management and provenance infrastructure throughout several countries with a strong e-government background (Ojo & Adebayo, 2017). To dive deeper into the context of blockchain technology in the public sector, the following section presents benefits and application scenarios in this very domain.

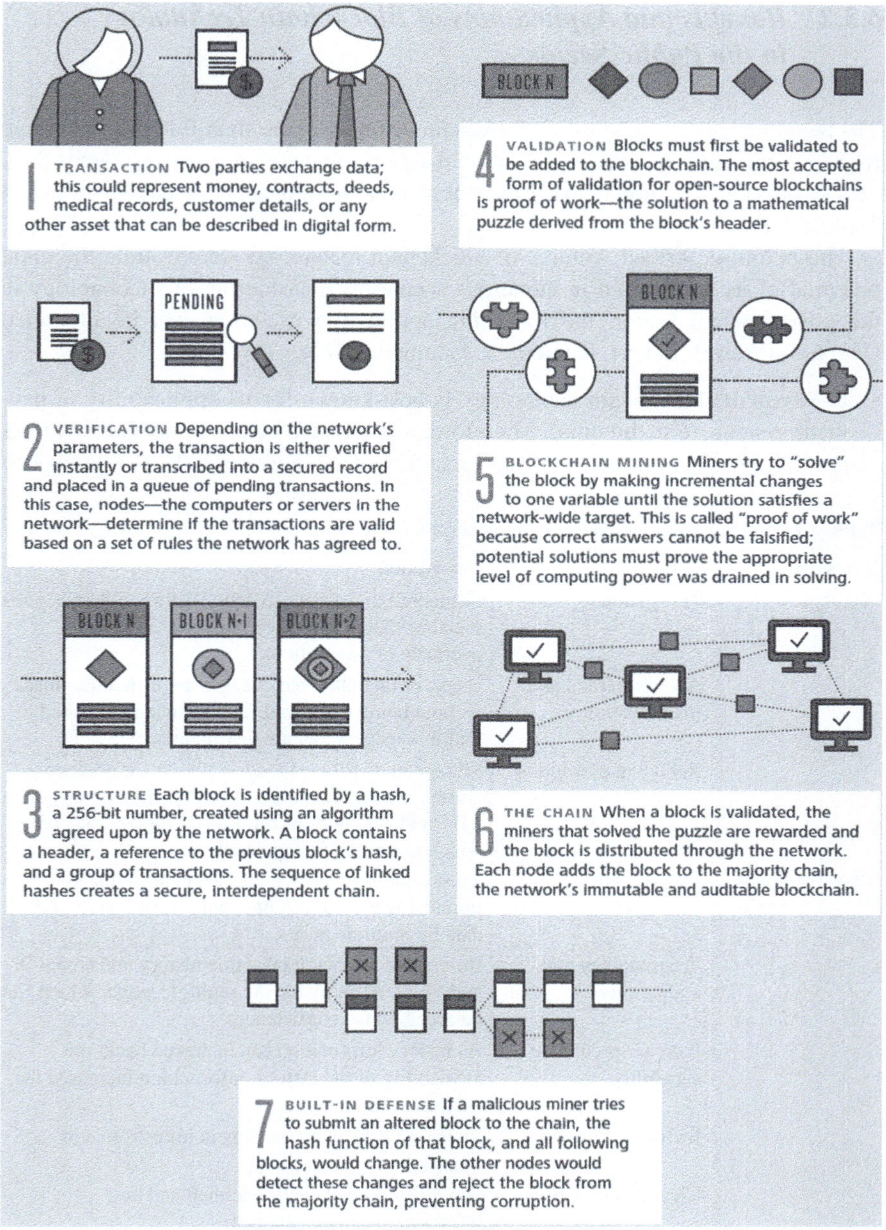

Fig. 6.2 The principles of a blockchain workflow. (Adapted from Piscini, Guastella, Rozman, and Nassim (2016))

6.3.2 Benefits and Applications of Blockchain Technology in the Public Sector

The benefits that can arise from blockchain technology are manifold and can range from strategic aspects, over organisational aspects, to economic aspects. Ølnes et al. (2017) provide a comprehensive overview over these aspects as can be found in Table 6.2.

The before-described features of blockchain technology demonstrate the great potential of its application in numerous scenarios. Considering this technology in the governmental sector, the following application use-cases can be identified (Fig. 6.3) (Welzel, Eckert, Kirstein, & Jacumeit, 2017):

- **E-Payment:** blockchain technology is best-known for its applicability in payment systems (e.g., bit coin). Therefore, it could also be used to make payments towards the government and vice versa. Examples here could be tax payments/

Table 6.2 Benefits and features of BC in governmental application

Category	Features	Description
Strategic	Transparency	Democratizing access to data. History of transactions remains visible and every node has complete overview of transactions.
	Avoiding fraud and manipulation	Hacks or unauthorized changes are difficult to made without being unnoticed, as information is stored in multiple ledgers that are distributed.
	Reducing corruption	Storage in distributed ledgers allows for preventing corruption. For example, by storing landownership in a BT and having clear rules for changing ownership which cannot be manipulated.
Organizational	Increased trust	Trust in in process by increased control due to immutable recordkeeping and by verification of the data by multiple nodes.
	Transparency and auditability	Being able to track transaction history and create an audit trail. Also, by having multiple ledger which can be accessed for consistency.
	Increase predictive capability	As history information can be traced back, this availability of the historic information increased the predictive capability.
	Increased control	Increased control by needing consensus to add transactions.
	Clear ownerships	Governance need clearly defined and how information can be changed.
Economical	Reduced costs	The costs of conducting and validating a transaction can be reduced as no human involved is needed.
	Increased resilience to spam and DDOS attacks	Higher levels of resilience and security reduces the costs of measure to prevent attacks

(continued)

Table 6.2 (continued)

Category	Features	Description
Informational	Data integrity and higher data quality	Information stored in a system corresponds to what is being represented in reality due to need for consensus voting when transacting and distributed nature. This result in higher data quality.
	Reducing human errors	Automatic transactions and controls reduces the making of errors by humans.
	Access to information	Information is stored at multiple place which can enhance the easy the access and speed of access.
	Privacy	User can be anonymous by providing encryption keys or access can be ensured to avoid others to view the information.
	Reliability	Data is stored at multiple places. Consensus mechanisms ensures that only information is changed when all relevant parties agrees.
Technological	Resilience	Resilient to malicious behaviour.
	Security	As data is stored in multiple databases using encryption manipulation is more difficult. Hacking them all at the same time is less likely.
	Persistency and irreversibility (immutable)	Once data has been written to a BC it is hard to change or delete it without noticing. Furthermore, the same data is stored in multiple ledgers.
	Reduced energy consumption	Energy consumption of the network is reduced by increased efficiency and transaction mechanisms.

Adopted from Ølnes et al. (2017)

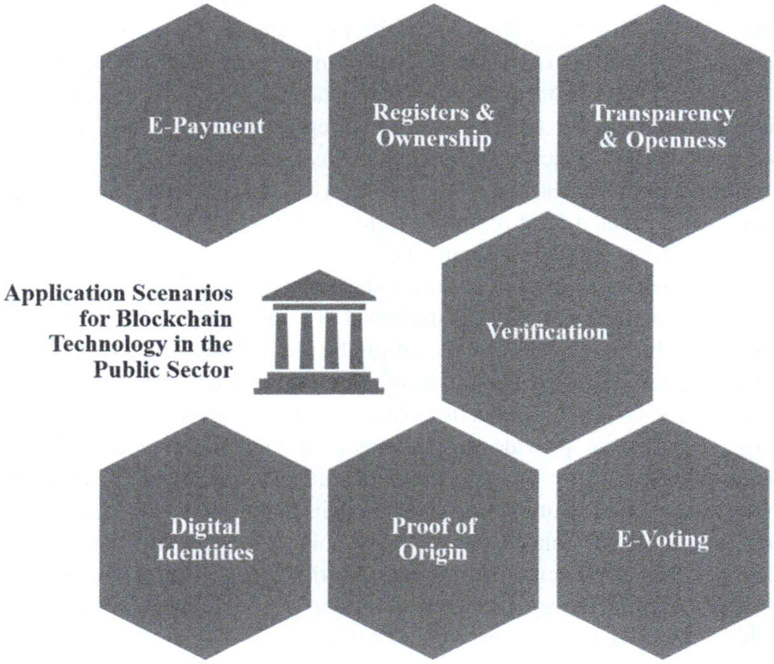

Fig. 6.3 Blockchain application scenarios. (Adapted from Welzel et al. (2017))

refunds, fees for certain services, as well as fines for violations. But not only monetary transfers between the citizens and the government, but also payments within the government as organization could be covered. These would include payment of salary, food stamps, parking tickets etc.

- **Registers and Ownership:** public registers, legal titles, as well as cadastres are common application examples for blockchain technology. The blockchain provides with its inherent transparency and immutability the means to prevent corruption, manipulation of existing entries, as well as a straightforward transfer of ownership. Furthermore, BCT can enable and enhance between governmental organizations on a national but also on an international level regarding the exchange of information, documents, and the verification of the existing of these documents.
- **Verification:** Verification of documents and data as well as their integrity are usually achieved via the use of digital signatures. This technology is established and currently used throughout different domains, including the governmental sector. Yet, they add an additional level of overhead to the process. First, there is the need for a central, trusted authority that issues the signatures and thus confirms the identity of the person acquiring the signature. Second, in order to be able to work with the signature, additional devices and/or software components are required, which add additional costs as well as might block certain application scenarios. BCT could help to reduce the burden of document verification and therefore increase the speed of the overall process.
- **Proof of Origin:** BCT can provide benefits in scenarios, where the traversal of a product through a process, e.g. a supply chain, has to be monitored in a way that every step can be verified. This can contribute to the fulfilment of legal compliance requirements. The public administration can also tap into this potential in cases, where it has the responsibility to govern over critical product/process flows, such as food chains or the trade with rare goods such as diamonds or art pieces.
- **Digital Identities:** the integrity of a digital representation of all ID relevant attributes can also be verified via a blockchain, by hashing all relevant attributes and storing the hash values within the chain. This concept could even be pushed further to use it as a kind of single-sign-on (SSO) system for organizations by including access rights to systems and services. The chain can then be used to check, if a person is allowed to access the particular service, system, or files. In addition, changes to the rights (withdrawal of rights or the addition of rights) can be seen via the history of changes within the blockchain.
- **Transparency and Openness:** today's society is demanding for transparency regarding the processes and actions taken by the government. Blockchain technology can help to provide this transparency and therefore contribute to the increase of overall trust of society towards its government and the elected representatives. A good example can be found in open data portals, which release open governmental data to the general public. By using BCT, the origin and integrity of this data can be verified, again improving trust towards the released information from the government, including accountability. Another example could be the budget of a government or parties, revealing all transactions and

spending, including donations and in consequence, making any lobbying activities and potential bias transparent.

- **E-Voting:** the matter of electronic voting is being discussed from various viewpoints, starting from e-voting being already implemented up to being completely anti-e-voting. Besides the principle "yes or no" discussion, BCT can be used for the voting process. Similar to the concept of bitcoin wallets, political candidates or parties could be equipped with a digital wallet and each citizen could vote with his or her specific single token towards the candidate or party. The candidate/party with the most tokens within their digital wallet win the election. While from a technical standpoint this is convenient, the approach also includes several caveats such as giving up to some degree anonymity of votes or could encourage tactical voting (as the number of votes are instantly visible) as well as potential bribing for securing the pivotal votes.

6.4 Real-World Examples of Open Data Infrastructures

6.4.1 Industrial Data Space

The German project Industrial Data Space (IDS) is one example of an open data infrastructure, with a particular focus on industrial applications. The IDS is based on the following core principles (Otto et al., 2016):

- **Data sovereignty**: the control over data within the IDS is never given up by the owner of the data. Thus, it is possible to link the data with licensing/terms and conditions that regulate operations with this data.
- **Secure data exchange**: a dedicated layer offers secure exchange of data between two or several entities, not only on a point-to-point bases, but also throughout complex supply chains.
- **Distributed architecture**: the IDS interconnects via its IDS connector all endpoints towards a distributed net of participants, without the necessity of a central authority or single-point-of-failure. The exact type of the architecture is set by the application scenario and is driven by economic aspects, specific to the market and domain at hand.
- **Data governance**: as described before, there is no central authority within the IDS. Therefore, participants of the IDS have to agree to a common rule set of how to work together, including duties and responsibilities. While this can be tricky to find common ground, at the same time, it provides the necessary flexibility to open the IDS for any application scenario and domain.
- **Network of platforms and services**: as the IDS is embracing the paradigm known as "Internet of Things" (IoT), the role of a Data Provider is not only limited to individuals or organizations, but can be also taken by devices, e.g., production machines, vehicles, etc. In additions, other Data Spaces/Markets can also interact with the IDS, and therefore with its entire ecosystem of stakeholders.

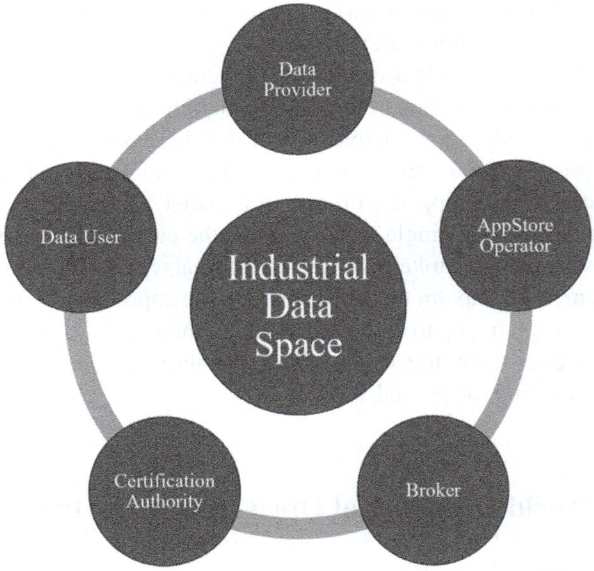

Fig. 6.4 Role concepts within the Industrial Data Space

- **Trust within the IDS**: without a common level of trust within the data space environment, participating actors will not engage with each other in terms of data exchange as well as service consumption. It is for this reason that participation is only possible by using the IDS connector, providing the required means of authentication and authorization.

While main goal of the IDS is the facilitation exchange between Data Providers and Data Users, other actors take important roles within this facilitation process (see Fig. 6.4). The actor environment within the IDS allows for a participant to enact several roles, including the possibility to rely on third parties for fulfilling tasks on their behalf. In the following, the distinct roles and their function within the environment of the IDS are explained (Otto et al., 2016).

The **Data Provider** holds the access to the sources from which data is offered towards the other participants of the IDS, while the data provider always keeps the control over the data. Furthermore, it offers descriptive information for the Broker to be able to properly register the data and offer it to interested stakeholders/actors throughout the IDS. The Data Providers is also responsible for the entire processing of the data within the IDS, including required transformations according to the inherent data model of the IDS, along with any applying terms and conditions in regard to the data itself. Finally, the Data Provider also orchestrates requests of data, in conjunction with handling the entire app and service ecosystem of the IDS. The role of the **Data Users** within the IDS is based on the consumption of data and services/apps, provided by other actors (Data Providers). This can either be a single source or multiple sources, including the required transformational as well as mapping-based actions, which are required to achieved compatibility with the targeted data model.

The **Broker** functions as intermediary, bringing together the searching party (Data Users) with the providing party (Data Providers). Furthermore, the Broker acts as central register for data sources within the IDS. Thus, the Broker also handles services such as the provision of means for Data Providers to publish their data, as well as the provision of search and retrieval capabilities for the Data Users to browse the registered data sources. In consequence, the Broker also facilitates the creation of agreements and the associated provision of the data between the involved parties. The exchange of data is therefore supervised and recorded to ensure a secure and complete transaction. This also includes potential rollbacks in case that a transaction failed. As the Broker is a central role within the exchange of data, it can also be set up to offer supplementary services to all involved parties such as quality assessment of data, or additional analytical services. The **AppStore Operator** holds the central authority regarding 3rd-party software, developed by participants to be distributed within the digital business ecosystem of the IDS and its AppStore. Therefore, the AppStore Operator provides means of describing and registering software to be offered to customers, including the download of these services, as well as payment functionality and rating options for the offered software services. Finally, there is the **Certification Authority** which exists to ensure that all components on the IDS meet the jointly-define requirements of all participants. This includes activities such as the handling of the entire certification process, starting from the request up to the approval/denial of the certification, operation of the reporting system of testing parties, up to the issuing of actual certificates. To guarantee a consistent, fair, and comparable process, the Certification Authority maintains a criteria catalogue, which acts as basis for the certification process.

To demonstrate the feasibility of the concepts inherent to the IDS, the following use cases are developed and realized:

- **Truck and cargo management in inbound logistics:** supply chains often suffer from the fact that data is unnecessarily duplicated by involved companies, thus causing storage and synchronization issues between each particular stage of the chain. This results in higher costs due to increased processing and slower or even delayed delivery. Therefore, an increased level of transparency is required, enabling consistent monitoring throughout the entire supply chain and thus, improving transportation as well as quantitative and qualitative forecasts. A good example for the before-mentioned situation can be found in truck and cargo management. In order to guarantee an efficient and effective management process, it is crucial for all relevant data to be available once the truck arrives at its destination for follow-up tasks (e.g., check-in, job order planning). Yet, this data is not always available in a complete form, due to, e.g., different freight carriers employed by the shipping companies. The IDS will solve this issue by the introduction of suitable standards and a general simplification of the data exchange process (i.e., data regarding the order itself, data about the transportation such as GPS data, master data of suppliers).

- **Development of medical and pharmaceutical products:** for medical and clinical data being highly-sensitive due to its personal aspects, it is also highly-heterogenous, as it consists of data by individuals, institutions, and machines. Also, this kind of data is due to its sensitive character rarely aggregated within one single place. Thus, this fact also represents a hurdle within the process of developing new treatments, therapies, and medication. But availability alone is not enough, information about the context of the collection process, as well as the involved IT systems, and the overall quality of the data itself are imperative to generate a complete picture. To overcome these issues, IDS will provide means for aggregation of data, as well as the required transformations to enable analyses. This will not only strengthen ongoing studies, but also allow for hypotheses testing beyond existing scales and flexibility. Combined with the open and standardized interfaces of IDS, various systems can be interconnected to enhance processing, visualization, and exploration of data. Furthermore, anonymization services will provide the requirements defined by law to fully-comply with GDPR and associated laws and regulations to protect privacy.
- **Collaborative production facility management:** modern production environments require a high level of data completeness, e.g., regarding individual components, utilization of machines, material availability etc. Currently, costs regarding the collection, analysis, and distribution of this data are high, as often this data cannot be collected by the companies own capabilities and therefore requires 3rd party support. While developments regarding standardized BUS-based systems have improved over the years, interconnectivity and data exchange represent still challenging tasks, especially while the entire sector faces an intense and ongoing transformation due to the Internet of Things (IoT) paradigm. This becomes even more obvious when considering the task of transferring sensitive company data towards the company's own perimeter. The IDS can step in at this point to act as a pivotal point regarding the secure and standardized exchange of data between different parties, especially across organizational borders. Furthermore, IDS can provide companies acting as participants additional services, which can support the companies in performing analyses on their data, which they have not been able to do before. Finally, manufacturers could offer their data on the IDS as well, opening new business models as well as to establish grounds for new cooperation between participants.
- **End-to-end monitoring of goods during transportation:** in certain domain, transportation of highly-critical goods requires for special monitoring during the transportation process to avoid damage or destruction of the goods themselves. Examples can be found in form of electronics, medical supplies, or chemicals. These damages can occur due to high and/or rapid temperature changes, shock/vibration, light exposure etc. Potential countermeasure come with, e.g., sensors that can communicated changes in the environment the goods are currently traveling, or the status and condition of the goods themselves. The IDS enables secure and complete end-to-end monitoring of the

transportation, informing customers and suppliers alike in case something should be wrong with the goods. The IDS therefore covers an important aspect of future IoT applications.

6.4.2 Data Market Austria

As the volume of data in our today' society is growing by the minute, it is more than natural for it to be considered an important "raw material" throughout all industrial and business sectors alike. In consequence, an effective and efficient ecoservice for handling this data within the Austrian economy is an imperative factor for sustainability regarding business and society as a whole. Currently, there is no agenda regarding such an ecosystem for Austria and ongoing initiatives are still working towards a significant breakthrough. While platforms regarding, e.g., governmental open data and open data from business exist, they are not connected, and business use cases have no common platform as a central host around this data. Yet, even with the data available, it often lacks a common data quality standard and thus suffers from interoperability issues. Therefore, the Data Market Austria (DMA) is trying to overcome these issues by performing the following actions[1]:

- **Advancing Technology Foundations**: this roadmap foresees three distinct steps. In the first one, blockchain technologies is used to incorporate a decentralized way of security of data registration, computation, as well as provenance. The second step builds upon brokerage services, including the use of sophisticated recommendation algorithms for an improved match of users and data/service providers. The third step ensures the provision of all required timely computational capabilities for all operations on the market, including the fusion of different data sources.
- **Creating a Data Innovation Environment**: DMA strives for the inclusion of various stakeholder groups, starting from start-ups and SMEs, large enterprises, academia, up to public administration. This will build an interaction and innovative environment on a co-creation basis, which allows for a variety of business models, guaranteeing the flexibility to provide long-term sustainable solutions for all involved parties.
- **Cloud-based infrastructure**: the DMA will host its services in a cloud environment, thus providing a transparent and highly-scalable service infrastructure for all participants and their individual use cases, applications and business models.

The Data Market Austria envisions similar roles as the Industrial Data Space. An overview of the seven different roles can be seen in Fig. 6.5. One of the most significant difference of the DMA and IDS is that DMA is not mainly focusing on the industrial sector, but aims to bring together stakeholders of different domains, sizes of companies as well as public administrations and actors from academia.

[1] https://datamarket.at/ueber-dma/

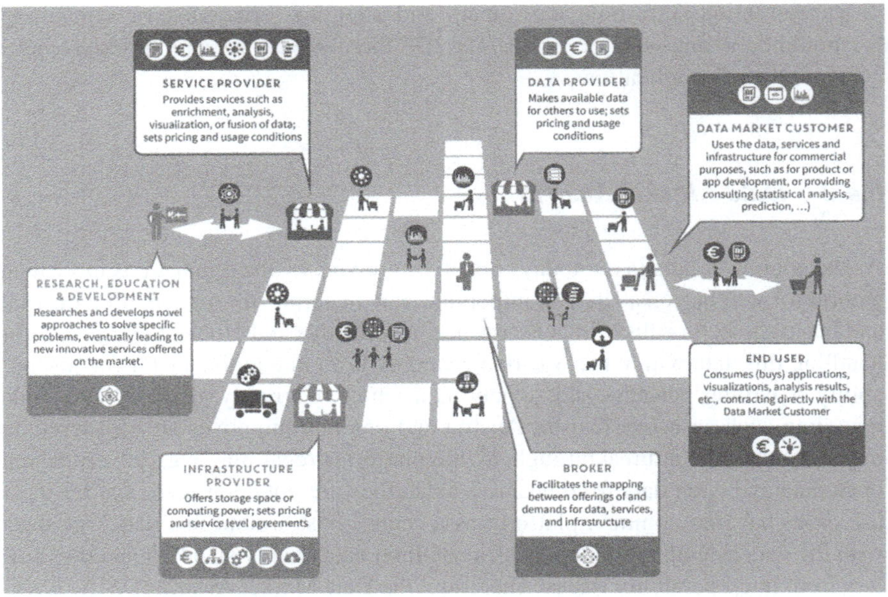

Fig. 6.5 Stakeholders within the Data Market Austria. (https://datamarket.at/en/)

To test the newly developed infrastructure, not only regarding technological requirements, but also in consideration of different business aspects of different domains, two main use case scenarios are covered within the DMA:

The first use-case is dedicated to the field of **Earth Observation**[2]. Here, data providers shift their access schemata more and more towards open access. An example in the European realm can be found with the Sentinel missions. The European Space Agency (ESA) and the European Commission are following a new set of policies, providing free access to this satellite data for any entity interested. Thus, earth observation becomes more and more accessible to the public as well as the industry sector for being used in business cases. Yet, single access to information sources is often only partly covering the requirements of Data Users, and linkage of data of different Data Providers is of great importance. For this reason, the DMA foresees Earth Observation data and services to be hosted on a cloud platform, allowing user to share infrastructure and data alike. This is in-line with the ESA's current activities regarding the process of distribution of Sentinel satellite data via a network of distributed data hubs. The DMA will demonstrate its capabilities in the EO domain, along with the linking of geodata on several application scenarios in the area of forestry.

The second use-case is placed in the **Mobility**[3] sector. Current data storage solutions are not suited for processing millions of data events transmitted, as they

[2] https://datamarket.at/earth-observation/

[3] https://datamarket.at/mobility/

can occur within, e.g., an IoT environment. Thus, a high level of scalability is imperative for future industrial but also other related business cases. Connected mobility solutions do present such as business case and application domain. Within this area, real-time prediction is considered one of the most time-consuming and computational demanding tasks. Thus, DMA will demonstrate its feasibility on two application examples in this field. The first example is dedicated to the task of **Taxi Fleet Management**. Here, public data and proprietary data will be used to optimize planning of taxi placement. Examples for data to be used are public transportation data such as arrival times of planes and trains, weather forecasts, local events, and mobile phone data of users that opted-in to make this data available. The second example comes in form of **Historical Traffic Flow Characteristics**. It is intended to derive patterns from historical data regarding traffic flow and mobility preferences of customers. This kind of data and predication could not only be of interest to taxi fleets but also towards city and urban planners to optimize traffic concepts as well as other related process towards improved traffic characteristics of the entire city.

6.5 Conclusion

In this chapter we have discussed the importance of open data infrastructures for a society, from both perspectives, the economic perspective as well as the governmental perspective. We have seen the high level of functional requirements that have to be fulfilled, in order to develop a sustainable infrastructure for open data and data in general. One of the most important aspects is present in the requirement of transparency and trust of the citizens towards the infrastructure as well as of the governmental organisations towards their potential users. State-of-the-Art technologies such as blockchains can help to provide the required level of transparency, while being open towards a variety of use-cases. We have discussed several use-cases in the domain of public administration and have seen that some of these could be realized already today, while for others it has still to be seen, if they can survive the scepticism of all involved parties as well as current legal obligations. While maybe in the public sector its advent is still not quite there, for sure it is becoming more and more present in the economic domain. In combination with open data, this has the potential for a huge variety of profitable business models. For more information on the value chain of open data and associated business models, please continue towards Chap. 7.

Chapter 7
Open Data Value and Business Models

"Open data is the new oil of the digital economy."

7.1 Introduction

The chapter focuses on innovation processes aspiring to generate value through a purposeful and effective exploitation of data released in an open format. On the one hand, such processes represent a great opportunity for private and public organizations while, on the other, they pose a number of challenges having to do with creating the technical, legal and procedural preconditions as well as identifying appropriate business models that may guarantee the long term financial viability of such activities. As a matter of fact, while information sharing is widely recognized as a value multiplier, the release of information in an open data format through creative common licenses generates information-based common goods characterized by nonrivalry and nonexcludability in fruition. An aspect posing significant challenges for the pursuit of sustainable competitive advantages.

The objective of this chapter is to shed light on some of the challenges highlighted above, with particular reference to the business models that may be adopted for igniting data-driven value generation activities. More specifically, the chapter will start by providing some background on a few key concepts having to do with the notion of value, the economics of information and business models. Subsequently, an overview of the most prominent studies on business models for open data will be presented. Finally, the main exploitation opportunities and some real-life cases will be discussed to exemplify a number of good practices of open data valorization in both the private and the public sector.

Y. Charalabidis et al., *The World of Open Data*, Public Administration and Information Technology 28, https://doi.org/10.1007/978-3-319-90850-2_7

7.2 Key Concepts

The discussion conducted in the following sections will address the value of open data and the different exploitation avenues that may be pursued from both a public and private perspective. The brief review presented in this section will thus glimpse at three concepts that are at the heart of open data exploitation processes: the notion of value, the cost structure of information and the concept of business model. The aim of this section is thus to create a clear and shared understanding to be used as a starting point for further discussion.

7.2.1 Value

As Adam Smith (1776) reminds us, when talking from an economist's perspective "the word value has two different meanings, and sometimes expresses the utility of some particular object, and sometimes the power of purchasing other goods which the possession of that object conveys. The one may be called 'value in use'; the other, 'value in exchange'. The things which have the greatest value in use have frequently little or no value in exchange; on the contrary, those which have the greatest value in exchange have frequently little or no value in use".

When taking a philosophical stance, traditional axiology shows how it is possible to distinguish between intrinsic value and instrumental value. In other words: if something is good only because it is related to something else, then its value is instrumental to the achievement of a given objective. To exemplify, money is supposed to be good, but not intrinsically good: it is supposed to be good because it leads to other good things such as the possibility to buy food and water (Schroeder, 2008).

In addition, the so called point of view theory (Schroeder, 2008) clarifies the difference between what is good simpliciter from what is good for a specific stakeholder: the former defines what has value from a more generic point regardless of the circumstances, while the latter is perspective-dependent.

Finally, the perception of value is strictly correlated with the needs of a society. In this respect, it is useful to mention that individual as well as collective needs may be hierarchically organized in order to provide a priority ranking. The work conducted at the beginning of the last century by the American psychologist Abraham Maslow represents a cornerstone in this field (Maslow, 1943). His celebrated hierarchy of needs identifies five categories of needs having to do with physiology, security, belonging, esteem and self-actualization. In a resource constrained situation, such classification represents a useful tool in identifying and prioritizing the long term strategic priorities that should be targeted in order to create value for the society. A value that – as Savitz (2006) reminds us – unfolds along a number of dimensions touching upon financial, social, and environmental aspects.

7.2.2 Public Value

Moving on to the concept of public value, it may be described as the analogue of the desire to maximize shareholder value in the private sector: in fact, according to Kelly, Mulgan, and Muers (2002), all governments should want to maximize "public value added", i.e., the benefits of government action when weighed against the costs (including the opportunity costs of the resources involved). In addition, the notion of public value spawned the development of performance measurement/management frameworks, attracting the attention of practitioners and management enthusiasts.

Taking this stance, Kelly et al. (2002) discuss public value as an analytic framework for public sector reform where public value becomes "the value created by government through services, laws, regulations and other actions" thereby creating a "rough yardstick against which to gauge the performance of policies and public institutions". Cole and Parston (2006) crafted the Accenture Public Service Value Model's methodology for measuring how well an organization achieves outcomes and cost-effectiveness over a period of years and, adopting a sectorial perspective, Cresswell, Burke, and Pardo (2006) outlined a public value framework for the return on investment (ROI) analysis of government IT estate. Despite some difficulties in operationalizing the concept through wide-ranging measurement systems, the notion of public value may offer a promising way of measuring government performance and guiding policy decisions.

7.2.3 Business Model

The notion of value is at the heart of business models. They have been integral to trading and economic behaviour since pre-classic times (Teece, 2010) nevertheless, the business model concept became prominent with the advent of the Internet in the 1990s and it has been gathering momentum since then. As it often happens in the academic field, no consensus has been reached on a common definition for such concept. The literature, in fact, refers to a business model as a statement (Stewart & Zhao, 2000), a description (Applegate, 2000; Weill & Vitale, 2001), a representation (Morris, Schindehutte, & Allen, 2005; Shafer, Smith, & Linder, 2005), an architecture (Dubosson-Torbay, Osterwalder, & Pigneur, 2002), a conceptual tool (Osterwalder, 2004; Teece, 2010) a structural template (Amit & Zott, 2002), a method (Afuah & Tucci, 2002), a framework (Afuah, 2004), a pattern (Brousseau & Penard, 2006) and as a set (Seelos & Mair, 2007).

For the purpose of the present discussion, the notion of business model will be intended as a representation of the value architecture through which a given enterprise generates, delivers and appropriates value (Osterwalder & Pigneur, 2010). Business models thus provide an enterprise centric view and are tightly connected with the notion of value. Specifically, the key challenge that we will be discussing in this chapter is the identification of the value architectures (business models) that may be put in place for the generation of both public and private value.

In order to properly design financially sustainable and strategically cunning business models it is important to have a deep understanding of the economics of information and its impact on exploitation strategies. In this respect one of the most authoritative contribution on the topic present in the literature has been offered by Carl Shapiro and Hal Varian in "Information rules: a strategic guide to the network economy" (1999). In their bestselling book, the two authors provide a clear and detailed account of the cost structure of information in terms of production, reproduction and distribution.

According to Shapiro and Varian (1999), one of the most fundamental features of information goods is that their cost of production is dominated by the "first-copy costs". Once the first copy of a DVD or an MP4 file has been generated, the cost of producing additional units is very low. In addition, the cost of distributing information is falling, causing first-copy costs to comprise an even greater fraction of total costs to get an information good in the hands of the final consumer. In the language of economics, the fixed costs of production are large, but the variable costs of reproduction are small. This cost structure leads to substantial economies of scale: the more you produce, the lower your average cost of production. But there's more to it than just economies of scale: the fixed costs and the variable costs of producing information each have a special structure. The dominant component of the fixed costs of producing information are sunk costs, costs that are not recoverable if production is halted. If you invest in a new office building and you decide you do not need it, you can recover part of your costs by selling the building. But if your film or your song flops, there isn't much of a resale market for its script or score. Sunk costs generally have to be paid up front, be/one commencing production. In addition to the first-copy sunk costs, marketing and promotion costs loom large for most information goods. The variable costs of information production also have an unusual structure: the cost of producing an additional copy typically does not increase, even if a great many copies are made. Unlike Airbus, Google does not face appreciable and lasting capacity constraints. Normally there are no natural limits to the production of additional copies of information: if you can serve one customer you can serve a million customers at roughly the same unit cost. The low variable cost of information goods offers great marketing opportunities. Just as sellers of new brands of toothpaste distribute free samples via direct mail campaigns, sellers of information goods can distribute free samples via the Internet. The toothpaste vendor may pay a dollar or two per consumer in production, packaging, and distribution to promote its product; but the information vendor pays essentially nothing to distribute an additional free copy.

The first-copy costs common to information goods are "merely" the extreme version of what we see in other industries where scale economies are powerful, which includes many high technology industries like chip fabrication. To summarise the brief overview on the economics of information, we may say that:

- Information is costly to produce but cheap to reproduce.
- Once the first copy of an information good has been produced, most costs are sunk and cannot be recovered.
- Multiple copies can be produced at roughly constant per-unit costs.
- There are no natural capacity limits for additional copies.

The cost structure of information goods is a key aspect to keep in mind when designing economically sustainable (and profitable) products or services leveraging open data as a strategic resource.

7.3 Open Data Value Chain and Business Models

The process that leads from the generation of a data asset to its consumption is far from being linear and subject to diverse interpretations. Many studies have embarked in providing a high-level representation of such process (Capgemini, 2015; DG Connect, 2013; Ferro & Osella, 2011; Pira International, 2010). The various attempts provided representations at different levels of granularity and units of analysis. For the purposes of this discussion a revisited version of the value chain proposed by Ferro and Osella (2011) will be used in order to include information generated both by public and for-profit actors as well as to clearly distinguish three aspects: (1) activities conducted, (2) relevant actors and (3) outputs generated in each step of the value chain.

As it may be noticed from Fig. 7.1, the main added-value activities conducted along the chain are: data generation, dissemination, retrieval, storage, categorization, exposure, re-use and consumption; while the outputs of the different steps are: raw data, refined data, and "fit-for-purpose" products and services; finally, 11 of archetypical actors (four public and six for-profit) operate along the value chain.

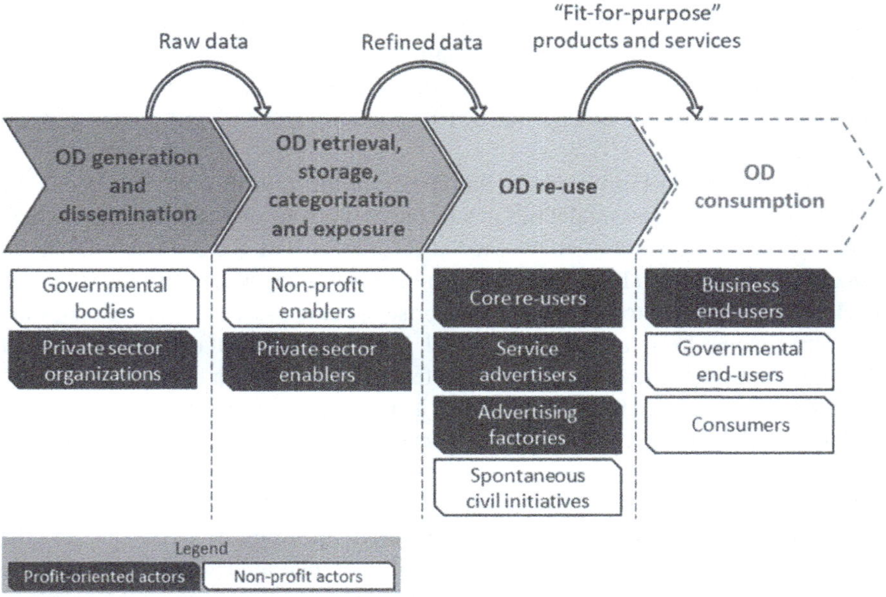

Fig. 7.1 Open data value chain. (Elaborated from: Ferro and Osella (2011))

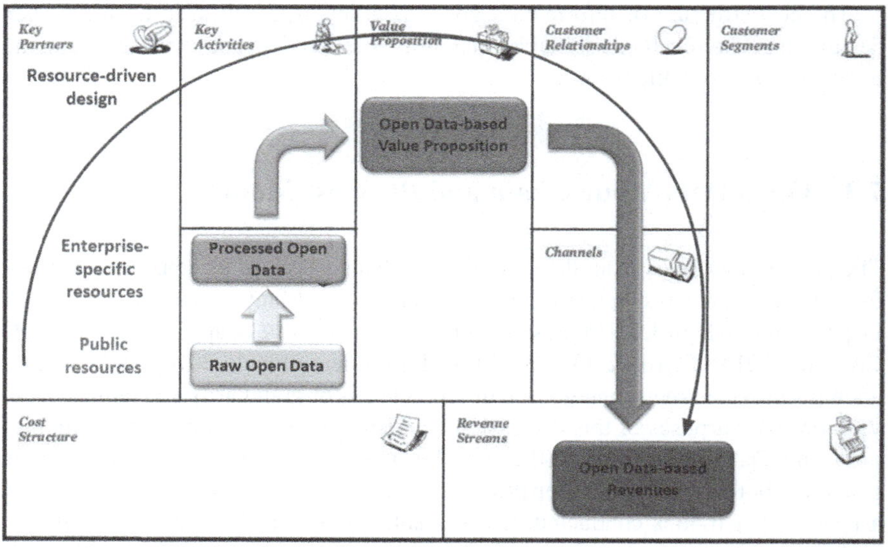

Fig. 7.2 A resource-driven design. (Source: Ferro and Osella (2011))

The discussion about which business models may be adopted in the exploitation of open data mainly applies to private for-profit organizations as they are the actors more challenged by finding a financial sustainability in leveraging a public good. It is important to underline that such discussion does not merely offer a representation of the activities conducted or the position covered in the value chain. As a matter of fact, to provide actionable insights to a would-be open data entrepreneur it is essential to depict the value architecture through which an organization creates, delivers and appropriates value. For this reason, the business model canvas methodology devised by Osterwalder & Pigneur, 2010 represents a useful and comprehensive tool (Fig. 7.2).

As highlighted in Ferro and Osella (2012), in the case of open data reuse the epicenter of the business model lies in a resource (i.e., one or many data sets) which is accessible by everyone when released in accordance with the open data paradigm (i.e., without technical, legal and price barriers). Subsequently, such a raw resource is elaborated in order to become an enterprise-specific asset that distinguishes the respective owner from the rest of the world. Such processed data is an ingredient of the value proposition that the enterprise offers to the market. In other words, elaborated data is "packaged" and embedded in the bundle of products and services which is supposed to create value for at least one customer segment. In return to such a value, customers generate revenues for the enterprise through alternative forms of payment. The discussion about business models employable in the exploitation of open data will focus on for-profit actors operating in the second and the third step of the value chain. More specifically, on two archetypal actors directly facing the end consumer (core-users and service advertisers) and two operating behind the front lines (enablers and advertising factories). For each archetype one or

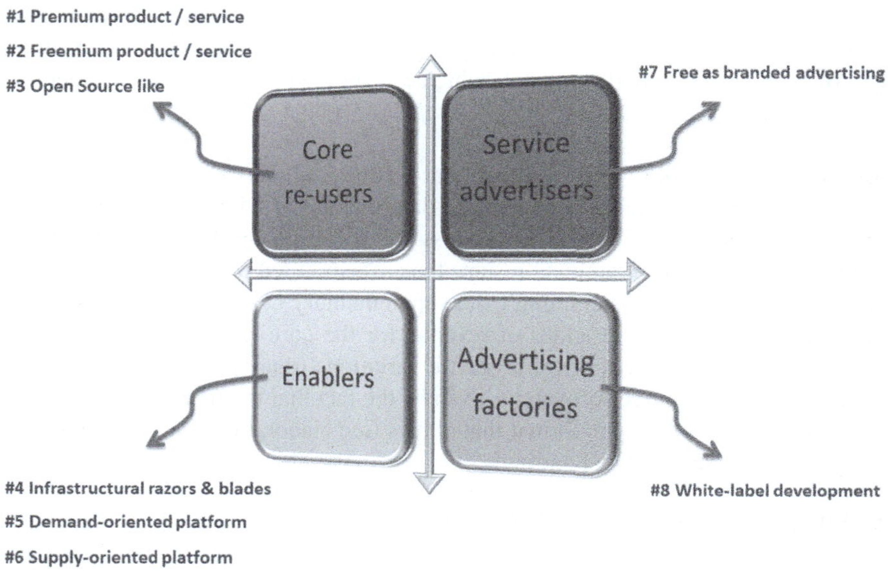

Fig. 7.3 Archetypal actors & business models. (Source: Ferro and Osella (2011))

more potential business model was identified and briefly described in natural language. A more formal representation of such business models may be found in Ferro and Osella (2013) (Fig. 7.3).

#1 Premium Product/Service While implementing this business model, a core re-user offers to end-users a product or a service presumably characterized by high intrinsic value in exchange for a payment that could occur à la carte or in the guise of a recurring fee: while the former implies the payment of an amount of money for each unit of product purchased (pay-per-use), the latter has an "all-inclusive" nature since it grants for a given timeframe the access to certain features in accordance with contractual terms. In this business model, probably associated to the "mainstream" model by the majority of analysts, the high intrinsic value, coupled with the price mechanism, calls for B2B customers often called "high-end market" (De Vries et al., 2011) and for long or medium terms relationships going beyond single transactions.

#2 Freemium Product/Service Core re-users resorting to this business model offer to end-users a product or a service in accordance with freemium price logic: one of the offerings is free-of-charge and entails only basic features, while customers willing to take advantage of refined features or add-ons are charged. In the open data realm, the implementation of this business model has its roots in limitations deliberately imposed by the core re-user in terms of data access: as a result, ad-hoc payments may be required to enjoy advanced features, to have recourse to additional formats or, sometimes, to weed out advertising. In contrast with the previous model, here the prominent target market is the consumer one often called "low-end market" (ibid) with which the firm establishes medium or short terms relationships that usually do not involve the customization. Target customers are generally reached via the Web or via the mobile channel, which are promising to "hit" a considerable number of installed bases.

#3 Open Source Like This very peculiar business model takes place on top of products, services, or simple unpackaged data that are provided for free and in an open format. In terms of economics, a cross-subsidization (Anderson, 2009) occurs in the enterprise under examination since the costs incurred for free offering of data are covered by revenues stemming from supplementary business lines that are still open-data-based: in fact, trickles of revenue for the core re-users may stem only from added-value services or from license variations (dual licensing). The resemblance with Open Source software is given by the fact that in this circumstance data is provided in a totally open format that allows free elaboration, usage and redistribution without any technical barrier.

#4 Infrastructural Razor & Blades Entering in the realm of enablers, this business model is chosen by enterprises acting as intermediaries that facilitate the access to open data resources by profit-oriented developers or scientists not driven by commercial intent. As it happens in the well-known model "razor & blades", the value proposition hinges on an attractive, inexpensive or free initial offer ("razor") that encourages continuing future purchases of follow-up items or services ("blades") that are usually consumables characterized by inelastic demand curve and high margins. Applying this model in the open data environment, datasets are stored for free on cloud computing platforms being accessible by everyone via APIs ("razor") while re-users are charged only for the computing power that they employ on-demand in as-a-service mode ("blades"). This business model exhibits another case of cross-subsidization whereby profits accrued from the provision of on-demand computing capacity cover costs attributable to the storage and maintenance of data. Finally, it goes without saying that application of this model is limited to contexts and domains in which the computational costs are significant.

#5 Demand-Oriented Platform Following this business model, the enabler acting as intermediary provides developers with easier access to open data resources that are stored on proprietary servers having high reliability. Once collected, open datasets are subsequently catalogued using metadata, harmonized in terms of formats and exposed through APIs, making it easier to dynamically retrieve data in a meaningful way. As a result, a wide range of critical issues pertaining to original raw data are made irrelevant due to the usage of platforms capable to convert datasets in data streams, contributing significantly to the "commoditization" and "democratization" of data. In addition, developers may reap the benefits given by the "one stop shopping" nature of such platforms: they may resort to one supplier and access a variety of information resources through standardized APIs – even beyond the borders of the open data – without having to worry about interfaces connecting to each original source. This "procurement" approach is crucial to minimize search costs and, by consequence, transaction costs. In terms of pricing, as a good that was born free and open (such as Open Government Data) cannot be charged in absence of added value on top of it, enablers adopting this business model earn revenues in exchange for advanced services and refined datasets or data flows. To sum up, re-users are charged according to a freemium pricing model that sets the boundary between free and premium in light of feature limitations.

#6 Supply-Oriented Platform To conclude with enablers, this business model entails the presence of an intermediary business actor having again an infrastructural role. However, on the contrary of the previous case, according to this logic open data holders are charged in lieu of developers. In fact, the enabler, following the golden rules of two-sided market (Eisenmann, Parker, & Van Alstyne, 2006) fixes the price according to the degree of positive externality that each side is able to exert on the other one. Consequently, this approach is beneficial for both sides of the resulting arena: from developers' perspective, their barriers are wiped out (i.e., they can retrieve data without incurring cost) while, from the governmental angle, open data holders become platform owners taking advantage of some handy features such as cloud storage, rapid upload of brand-new datasets by public employees, standardization of formats, tagging with metadata and, above all, automated external exposure of data via APIs and GUI. Public agencies that adhere to such programs in order to dip their toes into the water of Open Data establish long term relationships with providers and are required to pay a periodic fee that depends on the degree of sophistication characterizing the solutions purchased and on some technical parameters.

#7 Free as Branded Advertising Service advertising is an emerging form of communication aimed at encouraging or persuading an audience towards a brand or a company. Conversely to the more famous "display advertising", where commercial messages are simply visualized, in service advertising the advertiser strives to conquer the customer by providing him or her with services of general usefulness. That said, in the open data realm, services offered in this way do not generate any direct revenue, but they are supposed to bring positive return in a broad sense, driving economic results on other business lines – unrelated to open data – that represent the enterprise's core business. The rationale fueling this "enlightened" business model is twofold. Firstly, it may be based on a powerful advertising boost that leads the company to consider the cost as a promotional investment in the marketing mix. Secondly, it seems to be very convenient in presence of zero marginal costs (Anderson, 2009), a situation that occurs when the costs of distribution and usage are not significant.

#8 White-Label Development Last but not least, if service advertisers do not have in-house sufficient competencies required to develop their business endeavors, they can knock the door of advertising factories. Such firms, in fact, come into play as outsourcers carrying out duties that otherwise would be handled by service advertisers. Hence, the development of PSI-based solutions is particularly compelling for companies willing to use open data as "attraction tool" but not equipped with competencies required to do so (e.g., data retrieval, software development, service maintenance, marketing promotion). In order to let the service advertiser's brand stand out, solutions are developed in a white-label manner, i.e., shadowing the outsourcer's brand and giving full visibility to the sole service advertiser's brand. Taking into account the "one stop shopping supply" and the business-criticality of the solutions in terms of corporate image, the resulting one-to-one relationship between provider and customer is tailor-made and "cemented". Concerning financials, advertising factories collect lump-sum payments or recurring fees in exchange for turn-key solutions so developed, depending on whether the crafted solution takes the form of product or service: whilst in the former case service advertisers perceive the cost as CAPEX, in the latter one the respective cost assumes an OPEX nature.

To provide and clear and explicit link among: archetypal actors, business models and real life business ventures, some examples are provided in Table 7.1.

Although the table does not have any expectation of statistical representativeness or exhaustiveness, it is possible to note a concentration trend around few positions in the value chain. More specifically, the lack of market maturity seems to have led the majority of companies to either lean towards enabling open data fruition for third parties by helping public agencies to expose data sets in a machine-readable format or towards leveraging open data as a marketing attraction tool through the provision of branded value-added services free of charge.

The business models presented above are stemming from the results of the exploratory study conducted by Ferro and Osella (2013). Other attempts to shed light on the topic have been conducted by scholars and professionals around world with different slants and foci. To exemplify, Shuhaka and Tauberer (2012) looked into business models for the reuse of legislative data and identified a six business models mostly overlapping with those identified by Ferro & Osella (pay services (or premium), freemium, advertising, startup, crowdfunding, nonprofit, government).

Table 7.1 Examples of for-profit open data ventures

Archetypal actors	Business model	Companies
Core re-user	Preemium	HospitalRegisters
Core re-user	Freemium	Voglio il Ruolo
Core re-user	Open source-like	OpenCorporates, OpenPolis
Enablers	Infrastructural razor & blades	Public Data Sets on Amazon Web Service
Enablers	Demand-oriented platforms	Data Markets, Infochimps, Factual, SpazioDati
Enablers	Supply-oriented Platforms	Socrata, OpenDataSoft, Microsoft OGI
Service advertizers	Free as branded advertizing	IBM City Forward, IBM Many Eyes, Google Public Data Explorer
Advertizing factories	While-label development	Datamarketing

Source: Ferro and Osella (2011)

The work conducted by Suhaka and Tauberer looked at both for profit and nonprofit venture and took into consideration provisional business models as in the case of "startup" (a company operating on venture capitalists' funds). Another effort worth mentioning is that of Jennifer Tennison (2012) focusing on a number of pricing logics for open data that take inspiration from the open source world. More specifically, she identified the eight logics briefly explained below:

Cost Avoidance: may help organisations avoid the costs of Freedom of Information (FOI) requests. This applies only to data that is likely to be requested or has a very low publishing cost. Organisations that have a high FOI spend with lots of successful requests may find that they can lower that FOI spend by proactively releasing data (and making it easy to find).

Sponsorship: the reverse of cost avoidance is finding sponsors for open data publication. If there are people who strongly believe that a particular dataset should be open and available to all, they may be prepared to sponsor its publication (which is not the same as licensing it; the consequence is that the data is open for all, not just for those who pay). How to persuade others to sponsor opening up data? Perhaps, if it is the type of dataset that is hard to close up again after it has been made open, they might gamble that it would lower their long-term costs. Perhaps they sell analysis or visualisation products that they know those who use the data will find useful, and so getting the data available widely will aid their business.

Freemium: the freemium model has been used with some success for web-based services; it might also work for open data. Under this model, an organisation would publish open data in a basic form – perhaps with some limitations on formats and throttling of API calls – and offer advanced access to those who are willing to pay. There are many ways in which open data can be made more useful than static publication of spreadsheets or a basic API; under a freemium model some of these enhancements would only be offered to those who pay for it:

- availability of different machine-readable formats
- unconstrained numbers of API calls
- more sophisticated querying
- access to data dumps rather than through an API (or vice versa)
- provision of feeds of changes to the data
- enhancement of the data with additional information
- early access to data
- provision of data on DVDs or hard disks rather than over the net

Dual Licensing: data publishers could provide data under an open license for certain purposes, and under a closed license for others. This technique has worked for some open source products. The "certain purposes' might not be simply 'non-commercial": publishers could still encourage start-up use of the data by charging based on the size or revenue of the organisation. Or the license could state that the data can be used in products but cannot be used in further "added value" data feeds without being licensed (this is roughly equivalent to dual-licensing with a share-alike license).

Support and Services: offering support and services is a business model which seems to work well for companies built around open source. In the open data world, data publishers could offer paid packages with:

- guarantees on data availability
- prioritisation on bug fixes (both in data and its provision) for paying customers
- timely help for customers using the data
- services around data visualisation, analysis and mashing with other data

These kinds of services still tend to be coupled with licenses in the data world, whereas in open source they have been successfully disentangled.

Charging for Changes: in some cases, individuals or organisations are obliged to provide information to public bodies (and they have a statutory duty to collect it), so that it is available within government and more generally in society. These public bodies can (and sometimes do) charge the providers of that information "administration costs". Examples of this are Companies House information, the Gazettes, Land Registrations, VAT Registrations and so on. In these cases, those who supply the information to the register are bound to by law, so it would be possible to charge them whatever it took to support providing the data as open data. Indeed, supplying the data as open data is likely to increase its usage (both within government and more widely), and therefore the political pressure to retain the registry and thereby maintain its longevity.

Increasing Quality through Participation: the model used by legislation.gov.uk is based on increasing the quality of the data that we have to publish – bringing the statute book up to date – by enlisting the help of other parties who would benefit from having an up-to-date open statute book. Because otherwise this information is very costly to get hold of, there are any number of potential contributors, including publishers, lawyers, academics, and government itself. This model doesn't entirely cover the costs of opening up data: contributors are not generally paying money to be involved but donating effort to maintaining the published data. Thus, this business model does not completely cover costs, but it is a very useful one for organisations that have an obligation to publish information but lack the resources to do it well.

Supporting Primary Business: the final business model may be used when releasing open data naturally supports the primary business goal of the organisation. The best example of this is around the Barclays Cycle Hire in London, where releasing open data about the bikes drives the development of Apps that make it easier for potential customers to use the scheme, thus bringing in revenue to the core business. Another example is the recent release of data about Manchester City football players which, they hope, will lead people to create better ways of measuring player performance, which they will then be able to take advantage of.

A further, and final, perspective is offered by Janssen and Zuiderwijk (2014) who conducted a study on the business models for infomediaries, i.e.: organizations positioning themselves between open data producers and users. The authors identified six business models (single-purpose apps, interactive apps, information aggregators, comparison models, open data repositories, and service platforms) some of which describing the purpose of the tool developed and some others describing the activities conducted by the organizations building the tool.

As it may be noticed from the overview provided above, the topic of business models for open data exploitation still requires time and efforts to reach a maturity stage. As the availability and the quality of open data increase, it could be worth conducting a new wave of studies that go beyond mapping and formalizing business models by looking at their performance and long-term sustainability from a financial, legal and operational point of view.

In the following sections the discussion will shift from an enterprise centric view to a macro level perspective highlighting market and governance aspects that need to be addressed for the creation of a vibrant open data socioeconomic system.

7.4 Open Data Exploitation in the Private Sector

In order to understand what business opportunity may reside in the exploitation of open data for Europe-based enterprises, it is important to provide an overview of the estimated current market size, the expected growth trends as well as of the breakdown by sector and member state.

Figure 7.4 provides a quantification of the European market size together with some projections to 2020 along three main dimensions. The total market value was estimated to be close to 300 billion euros in 2016 and expected to more than double by 2020. The foreseen increase in value is reflected in a nonlinear fashion in the amount of companies that will integrate data in their core business as well as in workers whose main duties will revolve around data treatment or exploitation.

From a geographical point of view, the European data market follows a negative exponential distribution with a concentration of over 60% of the value in 4 of the 28 member states. Figure 7.5 shows the distribution among the different member states both in absolute value and in terms of percentage over the total.

Moving now from geography to industrial sectors it is interesting to note that the manufacturing and the financial sectors seem to offer the most significant opportunities. This may justify the significant attention that is currently being paid to topics such as industry 4.0 and fintech (Fig. 7.6).

With respect to the effort still necessary in terms of data liberation as a prerequisite for an effective and fruitful data exploitation, Fig. 7.7 shows that a long way is still lying ahead. A yearly survey conducted by the Open Data Barometer shows how at a global level only 10% of data is currently released in an open data format, thus significantly limiting the potential for reuse and exploitation. Having said that, of course, not all data should be released in an open format, especially those containing personal or sensitive information.

From a public-sector information standpoint, it is important to understand what could be the commercial appeal of the different datasets for private sector organizations in order to: prioritize investments in data liberation, allocate the resources

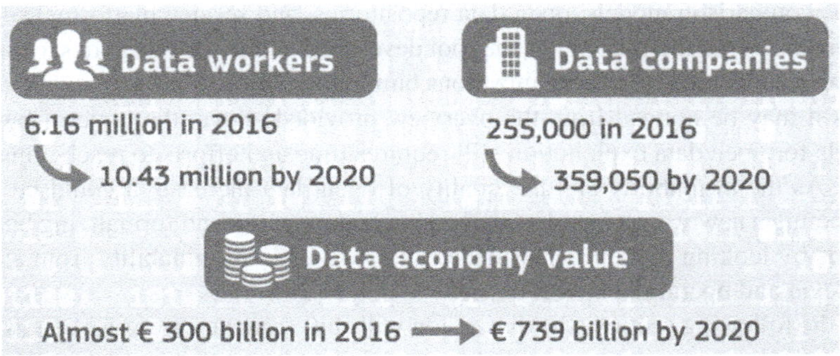

Fig. 7.4 EU data market overview. (Source: IDC (2017))

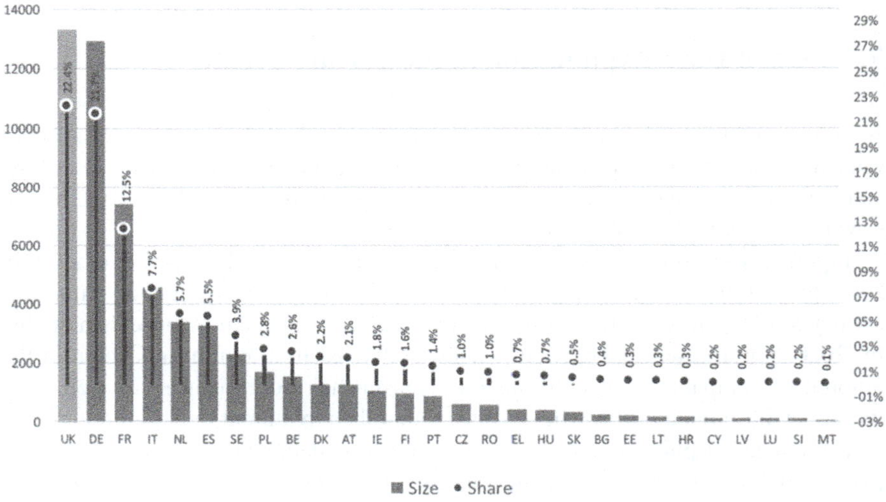

Fig. 7.5 Data Market Value (€M) & Share (%) by MS. (Source: IDC (2017))

necessary to guarantee the required levels of data quality and, finally, define a fair pricing model that may lead to a long-term sustainability of the process of data provision.

In this respect a study, conducted by Capgemini (2015) looked at the commercial reuse of open data sets. This study shows the different types of data generated by the public sector during its daily operations by appeal in terms of commercial reuse for profit-oriented business (see Fig. 7.8). Aside from noting that geographical together with meteorological and economic information seat of the podium of the classification, it is important to notice that not all data carry the same appeal and, as a consequence, should be exploited at the same time. This is to say that some data set are more readily reusable by the business ecosystem, while other types of datasets (e.g. cultural content) may require a longer lead time to find a viable exploitation avenue.

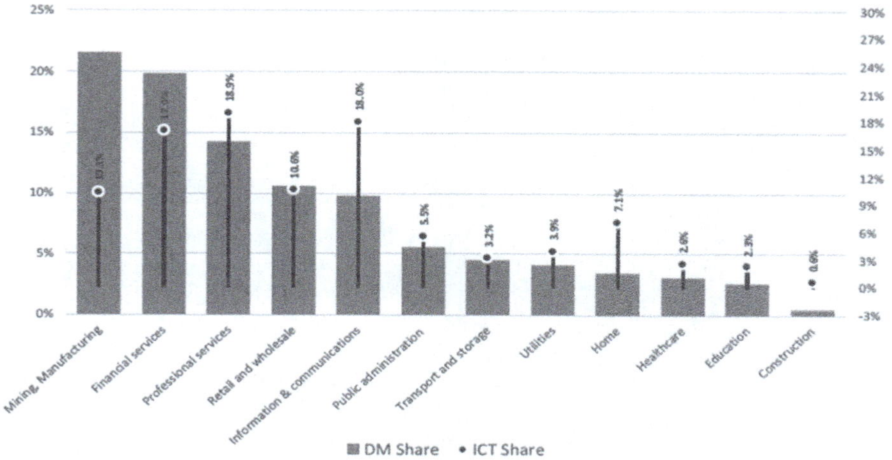

Fig. 7.6 Market size and ICT spending per sector. (Source: IDC (2017))

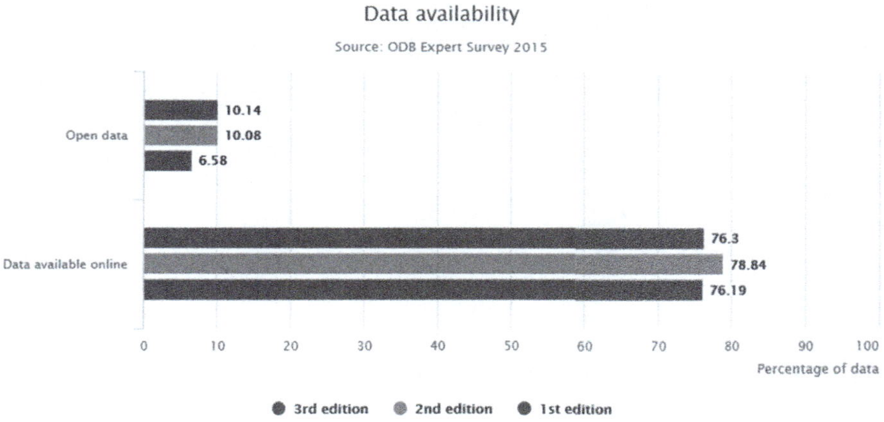

Fig. 7.7 Evolution of the availability of online data and open data. (Source: ODB (2016))

Finally, a strategic aspect to consider in the exploitation of open data as a key ingredient of a company product or service offering, is the identification of possible sources of competitive advantage necessary to consistently generate a performance that is superior to that of the other actors operating in the same competitive arena. When released in a fully open and reusable format, information may duly be considered a public good characterized by non-rivalry and non-excludability in consumption. As a consequence, the access to this type of resource may not be considered in itself a source of competitive advantage. Figure 7.9 shows how the focus in effort allocation shifts as a function of the degree of openness of the data sets exploited. In a situation in which legal, technological and price barriers are present, the company willing to exploit a given data set is required to spend significant resources in the process of data acquisition (especially for what concerns tech-

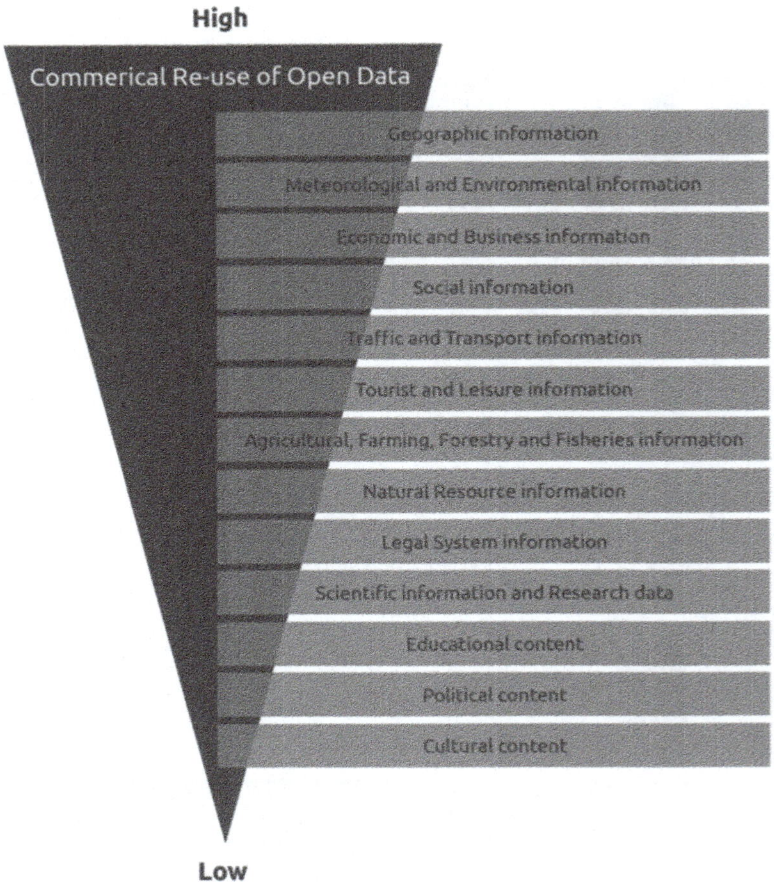

Fig. 7.8 Commercial reuse of open data. (Source: Capgemini (2015))

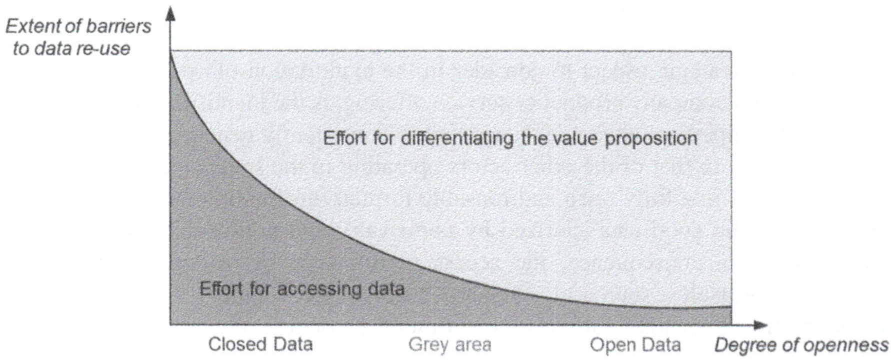

Fig. 7.9 Effort allocation as a function of data openness. (Source: Ferro and Osella (2011))

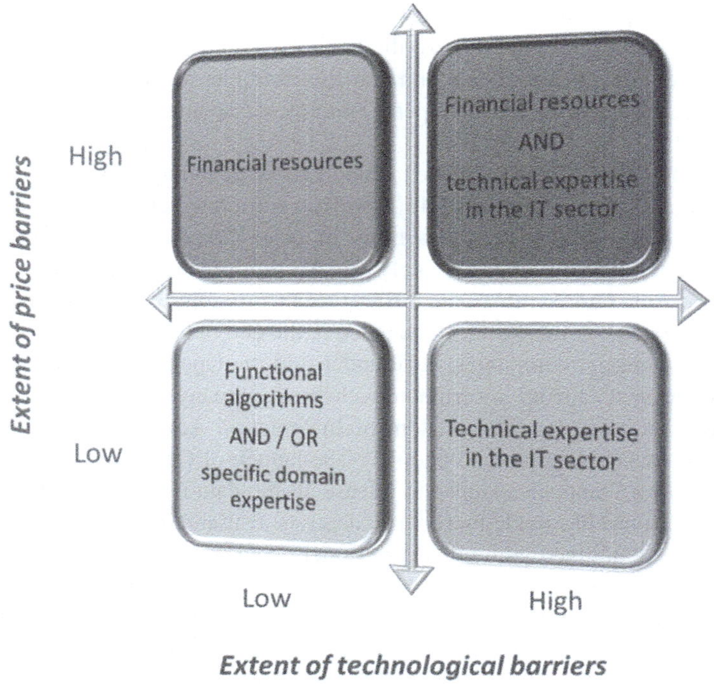

Fig. 7.10 Barriers and sources of competitive advantage. (Source: Ferro and Osella (2011))

nological and price barriers, as legal barrier may not be overcome). As the barriers to data re-use diminish, the focus of the company efforts moves from the process of data acquisition to the differentiation of its value proposition with respect to the competitors who, due to lower barriers to entry, increase in terms of numerosity.

The matrix depicted in Fig. 7.10 further clarifies the potential sources of competitive advantage that a company may exploit based on the presence and extent of price and technological barriers. When price barriers are significant and technological obstacles are negligible the availability of financial resources become the primary competitive edge discriminating between who can afford to access the information asset and who cannot. When, instead, technological barriers dominate over price barriers, technological skills become a must have to excel in the process of data acquisition, harmonization and integration. In contexts in which both type of barriers are present, the presence of both ingent financial resources and robust technological competences is required. Finally, when both price and technological barriers are not present or negligible, it is interesting to note that the sources of competitive advantage are no longer connected to the process of data acquisition, but rather they are related to functional algorithms for the treatment of data as well as to the presence of domain-specific expertise. While the former play a horizontal role and allow to differentiate the application logic of the service provided, the latter allow to contextualize the offering within a given vertical market.

In the final part of this section a use case will be presented and discussed in order to allow the reader to contextualize the knowledge and concepts presented in the previous sections into a practical and real-life example. More specifically, we will draw from and elaborate on the Open Corporates case study conducted by Becky Hogge (2016).

In 2010 the World Bank published a report showing that of 213 grand corruption investigations across 80 countries, 150 involved corporate vehicles that shielded the true beneficiaries of financial transactions. In these 150 cases, the total proceeds of corruption amounted to approximately $56.4 billion (Van de Does de Willebois, Halter, Harrison, Park, & Sharman, 2011). Open Corporates is the largest open database of companies in the world. It launched at the end of 2010 covering 3.8 million UK past and present companies. As founder told the Open Data Institute in 2012: "we take messy data from government websites, company registers, official filings and data released under the Freedom of Information Act, clean it up and using clever code make it available to people". The launch of Open Corporates predates the decision by Companies House to release all the data it holds as open data. But Companies House has made more basic datasets available for several years, and it was this data, combined with other government data sources (for example government spending data and Health and Safety notices) that fuelled Open Corporates in the beginning. Taking the same mixed input approach, Open Corporates has now expanded its coverage to over 105 jurisdictions and 85 million companies.

The added value that Open Corporates brings is the very detailed knowledge of how their database works. In addition, Open Corporates did "data-based advocacy", UK Department for Business were consulting on whether directors' and shareholders' full dates of birth should be published on the register, Open Corporates was able to demonstrate through real data that were dates of birth to be partially redacted, investigators would be unable to identify individual directors and shareholders robustly in cases numbering in the tens of thousands. OpenCorporates was also instrumental in pushing NGOs to demand the registry be made publicly available.

Open Corporates represents a very interesting case study in our discussion for a number of reasons: firstly, the business model they are implementing falls under the "open source-like" category identified by Ferro and Osella (2013) according to which the costs incurred for free offering of data are covered by revenues stemming from supplementary business lines that are still open-data-based. In this respect, considering that the whole Open Corporates database is freely available online and covered by an open license, the source of competitive advantage that the company may leverage to maintain its economic sustainability comes from a deep and detailed knowledge of the data base as well as of the domain. The second aspect of interest has to do with the fact that Open Corporates, not only acts as a open data advocate in the country in which they operate, but it helps breaking the silos present among public agencies working in countries both within and outside the European Union. Finally, Open Corporates may represent the dawn of a new paradigm in the pricing of data assets. More specifically, data released with an open license requiring any user to release derivative products in the same manner, may create the space for a new pricing logic that could require third parties to pay to maintain closed information assets generated by combining both closed and open data sources. This represents an inversion with respect to traditional pricing logics aimed at opening the access to information assets that could build on the diffusion of "open-by-default" as a mainstream approach as well as the diffusion of distributed ledger technologies like blockchain as an instrument to further promote transparency in the treatment of data.

7.5 Open Data Exploitation in the Public Sector

Shifting now the perspective from private sector actors to public agencies, this section intends to provide two contributions. The first has to do with the creation of a fully engaged and sustainable supply side, the second has to do with the investigation of the benefits that the public sector may enjoy as a savvier re-user of open data.

Despite the efforts put in place by an international and a highly motivated community of open data advocates operating from both within and outside the public sector, the "open-by-default" approach to date is still struggling to become a widespread practice and to generate the expected impact on the European socio-economic system. For this reason, there is an urgent need to take a new perspective on the topic in order to put cities, companies and citizens in the position to benefit from the significant, yet untapped, value residing in public sector's data vaults. More specifically, it is important to acknowledge the self-interested nature of human behavior by focusing on the benefits that public administrators may gain as stewards of government data vaults while viewing current drivers as significant, yet second order, positive externalities. Drawing on the principle that a thriving open data ecosystem requires the attainment of sustainability from the demand as well as from the supply side, the perspective proposed endorses governments' ROI as yardstick for gauging the ultimate feasibility of open data programs.

As a result, a new open data paradigm entails a radical shift in the way civil servants look at open data. This wave of change may be summarized as follows:

- From legal obligation to operational necessity
- From outward orientation to inward orientation
- From cost to opportunity
- From clerical function to strategic function
- From requiring a leap of faith to generating evidence-based impact

At an operational level, the implementation of such paradigm requires to rid of the "data liberation" approach in favor of an "open-by-design" principle allowing data to be born open through a revision of their generation process. This would represent a valuable tool in facing the challenges posed by a steadily growing pressure on public budgets. In addition it could contribute to make a further step towards the obtainment of an outcome-based government whose actions demonstrate a clear link with the results generated (i.e., outcomes) in terms of value that, in turn, could be internalized by the governments (e.g., efficiency, effectiveness) without overlooking the quest for the creation of value for society at large ("public value"). The adoption

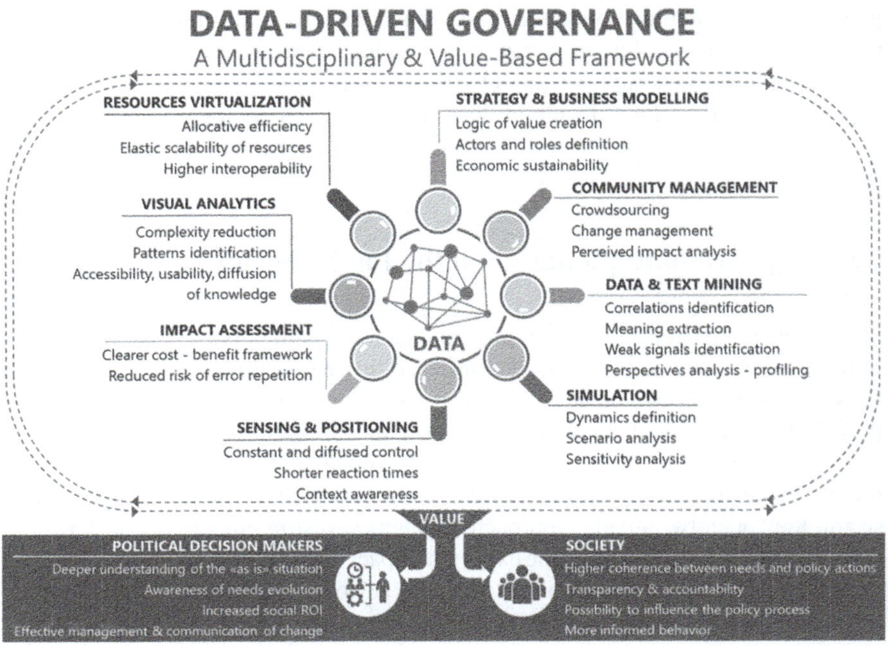

Fig. 7.11 Data-driven governance

of such an approach could represent a foundational step in the path leading to a data-driven governance paradigm briefly outlined in Fig. 7.11.

Placing data at the center of the governance process and combining it with a plurality of skills drawn from multiple knowledge domains represent the key ingredients for significantly improve the opportunities for value creation of a public decision maker. As a matter of fact, a data-driven multidisciplinary and value-oriented modus operandi may greatly benefit both decision makers and society at large. The former may gain a deeper understanding of the "as is" situation over which a given policy should be implemented to obtain a desired outcome, increase her awareness of evolution of needs to address, manage and communicate change more effectively and ultimately, increase the social ROI of any public investment. The latter, instead, may enjoy a higher level of alignment between perceived needs and policy responses, be more informed and incentivized to engage in the public debate thanks to higher levels of transparency and accountability. The creation of such virtuous cycle is believed to lead to a more effective and efficient allocation of taxpayers' money representing a key goal in times of shrinking public budgets.

To exemplify the benefits that the implementation of this approach may bring in terms of generation of value for society, a brief description of a use case conducted by OECD (2016) on the city of San Francisco is reported below. In the city of San Francisco, the heads of the foster care, juvenile probation and mental health departments, crafted an agreement with the city's attorney to permit the limited exchange of case information among agencies. The sharing enabled a new level of care for children interacting with any of these agencies. Case coordination improved, invisible populations emerged (overlapping clientele). This was made possible by the fact that the new integrated data system recognizes and focuses on the families that are most vulnerable, most troubled and most in need. Prior to data integration and data analysis the agencies had not realised that only 2000 users of services were using half of the resources of the department, and most of these families lived within walking distance.

As a follow up, the Human Service Agency concentrated delivery of services in specific neighbourhoods and co-located services at community centres, and this improved efficiency. Results included savings and better service delivery. Analysis of open linked data enabled a better assessment of needs of high risk youngsters diverting them from negative future events, the understanding of where youth were falling through, identification of what services were needed to intervene earlier and prevent negative outcomes. Initially supported by a low-tech system the solution was transferred to a more sophisticated platform to enable the three agencies to better understand the overlaps among their users. The crossover users of multiple systems were at higher risk of committing a crime (51% of San Franciscans involved in multiple systems were convicted of a serious crime, 1/3 had been served by the three agencies and 88% of these youths committed a crime 90 days after having become a crossover user – a critical window of opportunity for the case worker to intervene). A report produced highlighted a specific need: a web-based integrated case management system to make this connection in real time.

As services started being delivered by non-institutional care providers, the awareness grew of the need to balance the right of excellent care with the right to privacy protection. Hence, the need to carefully avoid sharing unneeded information. What made it so difficult where legal related matters. The preliminary good results convinced the district attorney's office that the integrated database could support better prevention services and gave the authorisation through a new statute that justifies the sharing of records on youth at particularly elevated risk levels. The school district decided to join to target students with high probability of dropping out to structure early intervention. Multi perspective on client's risk and identify protective factors. This can help agencies to determine which programmes are more effective, who needs to be targeted (most vulnerable, in trouble and in need) and how to coordinate the responsibilities. The San Francisco case study represents an excellent example of how a smarter exploitation of data by public agencies may lead to significant increases in performance.

7.6 Conclusions

The re-use of open data is believed to contribute to the world improvement for its potential to empower citizens, businesses, change how government performs, and improve the delivery of public services (Zeleti, Ojo, & Curry, 2014). The aim of the present chapter was to go beyond the glorification of the opportunities lying behind open data exploitation by exploring potential strategic viable choices from both a private and a public-sector perspective. Despite still being a phenomenon in its initial stages, the literature studying applicable business models to open data ventures offers some preliminary guidelines about possible strategic avenues that may be pursued in the design and implementation of potentially successful businesses leveraging open data. A portfolio of business models has been compiled as a toolkit from which would-be entrepreneurs or managers operating in established organizations may draw inspiration in the process of giving light to new companies or business lines. A reflection was also offered on the potential sources of competitive advantage may leverage in crafting their competitive strategy. As the barriers to data access decrease, it is possible to note a shift in the sources of competitive advantage for an organization. More specifically, the availability of financial resources and technical skills to be leveraged in the process of data acquisition becomes less relevant, while the presence of sophisticated functional algorithms and domain specific knowledge gains importance in the process of data elaboration and value extraction.

Shifting to a government perspective, a new approach to open data conceptualization and management in the public sector was proposed as a key complementary activity for the creation of flourishing open data ecosystem in which government agencies in addition to becoming reliable and efficient providers of quality data sets, become their first beneficiaries thus enabling a process of data-driven governance with significant positive spillovers for both policy makers and society at large.

Finally, to conclude the chapter, five synoptic principles are suggested to guide both public and private sector actors in a more purposeful valorisation of data assets. The principles are briefly described below:

- Size is not synonymous of value. That is to say, the assessment of data value should be based on a plurality of criteria: relevance for decision making, quality, and availability over time to name a few.
- Data science skills and the development of an evidence-based culture represents a key complementary ingredient to technological investments.
- Openness is a key driver of value multiplication. In other words, data should be released in formats maximizing the opportunities for the generation of economies of scope.
- Move beyond retrofitting. Rather than liberating data ex-post, the processes of data generation have to be open by design in order to minimize the cost of making them available to relevant stakeholders.
- Shared and clear values. The exploitation of data should be driven by shared values clearly identifying priorities in terms of advancing the environmental, social and economic conditions of the city.

The adoption of the above principles in the application of a long-term approach to data generation, exploitation and management may represents the necessary foundations to turn open data exploitation from a niche activity to a mainstream phenomenon as well as to make sure that the innovations contribute to the generated a positive impact on society in the quest towards the construction of a more sustainable and equitable world.

Chapter 8
Open Data Evaluation Models: Theory and Practice

"There is no unique model for open data evaluation. It depends on the perspective under evaluation."

8.1 Introduction

Evaluation of Open Data is a systematic determination of open data merit, worth and significance, using criteria governed by a set of standards (Farbey, Land, & Targett, 1999). It is an essential procedure trying to ignite a learning and innovation process leading to a more effective data exploitation. Examples of questions to be answered by open data evaluation could be: what is the current status of published data against the best practices identified, how effectively these data are published or used, what are the most valuable data for users, what are the problems and barriers discouraging the publication and use of open data and in which extend these barriers affects users' behaviour towards data usage. The answers on these questions will affect the next developments of an open data portal or initiative and the publication procedure.

A big challenge in the open data domain is how to evaluate open data in general and the platforms or infrastructures offering it and what are the metrics to be evaluated against to. For this reason, the value proposition of open data towards economic benefits for both governments and businesses and transparency for citizens has to be forecasted and evaluated. Different models and validation procedures have been used for the evaluation of open data and their provision portals examining different aspects of them. An aspect of evaluation could be the ability of both publishers and users to adopt and/or accept innovation or technology. Other aspects of evaluation could be the data maturity level or the quality of the published data. Another important aspect is the evaluation of impact originated and value created (net benefits) from the publication, use and reuse of open data. In order to assess those diverse aspects, several evaluation models and frameworks were developed in the domain of information systems.

We initially studied the developed evaluations models in the information systems domain providing insights about the targets of the evaluation procedure. Following these evaluation models, a first set of metrics and measures compiled targeting open

Y. Charalabidis et al., *The World of Open Data*, Public Administration and Information Technology 28, https://doi.org/10.1007/978-3-319-90850-2_8

data functionalities. As a next step, we were furthering our study to already developed metrics existing in the literature and classified them in specific categories. The main reason is the development of an overall assessment taxonomy, which includes every dimension of the quality of Open Data and their sources.

Following the "information system success" model, we are going to categorize different evaluation measures and benchmarks for the evaluation of data (Information Quality), platforms offering them (System Quality) and additional capabilities of those systems (Service Quality). Metrics for covering advanced functionalities based on the identified open data life cycle coming from various users (providers, users, pro-cumers) in Chap. 2 will also be demonstrated. In other words, the main objective throughout the chapter is to provide a classification of metrics, which could be used by public organizations and other stakeholders, in order to further develop evaluation models against different aspects of evaluation (readiness, impact and value creation, performance, quality, post-adoption etc.). The taxonomy aims at proposing various metrics, targeting different aspects of the evaluation: a public organization would then choose a different metric within the proposed taxonomy, according to each different aspect under assessment.

Furthermore, this chapter clarifies the distinction between the subjective and objective models for the evaluation of open data based on the identified evaluation models from the domain of Information Systems. Subjective are those models that concentrate on collecting users' opinions about a system towards the prediction of future behaviour or net benefits based on its perceived usefulness for the users. Objective models are those which are based on predefined metrics and values of them towards the assessment of specific benchmarks regarding the evaluated aspect (e.g. impact and readiness assessment).

The collected metrics could be used for the construction of both subjective and objective models regarding the utilisation of them in the formation of questions or the values space definition. For the subjective models, questions could be formed in order to ask users' opinions about a specific metric (to which extend does the system provide sufficient data?). For the same metric an absolute metric used in another model could be defined assigning values (<1000, 1000–100,000, >100,000 datasets) and searching for the answer in the platform under evaluation. Another example of absolute and quantities measurement is the percentage of completeness of a dataset (number of non-null values divided by the total number of all values) towards the assessment of its quality.

Both subjective and absolute metrics could be useful since they capture different views of the platform or infrastructure under evaluation. In the first case, the appraisal focuses on capturing the opinions of different types of users trying to assess in which extend they find the open data of their interest. The second case measures the values predefined metrics that could be used to categorise an open data platform based on its impact (low, medium, high) and/or maturity (allocating the platform under evaluation in one of the pre-defined maturity levels). It is worth to mention at this point that the metrics do not work alone, but in conjunction with other ones in order to reach a specific conclusion as it will be presented in the following sections.

Even more, subjective and/or objective metrics could be defined being part of the same evaluation model. Developing an evaluation framework, a researcher could utilise both subjective and objective metrics and measures. Finally, until now the presented models and examples falling in the category of quantitative research and evaluation. Qualitative methods could be used in order to capture unidentified aspects and difficulties in the domain of open data but using different techniques (interviews, SWOT analysis etc.). The qualitative methods could be used to generate questions based on the identified metrics towards revealing unknown problems, barriers and difficulties and getting deeper insights. An evaluation framework could utilise both quantitative and qualitative methods of assessment.

According to the above-mentioned objectives, the chapter consists of the following sections. Section 8.2 summarizes on basic background research in the domain of information systems evaluation models. It defines concepts, models and metrics used on Open Data and aims at both presenting the bibliographic research conducted on the issue and listing the criteria upon which the taxonomy/ analysis framework is later built. Section 8.3 presents applications of evaluation models in the open data domain while Sect. 8.4 compiles the evaluation metrics for open data in a taxonomy. Section 8.5 concludes the chapter and provides insights for further evaluation developments.

8.2 Evaluation Models in Information Systems

The scientific field of Open Data is very broad. In such a large problem space, the identification of focal points of assessment is essential. In general, when building an evaluation framework, a researcher decides on the aspect to evaluate and the model to use. The model could be either subjective or objective. Then she/he defines the problem space (functionality and/or quality) and poses the basic questions. The questions are posed according to the open data metrics, which will formulate the desired analysis framework. In this section, we provide the bibliographic background of the information systems evaluation models used for the evaluation of any information system, such as open data platforms and e-infrastructures.

For the development of any methodology we should take into account approaches and frameworks developed from **four subjective and quantitative relevant streams** of previous IS research on: (i) IS evaluation, (ii) IS acceptance, (iii) IS success and (iv) e-services evaluation. Additionally, several **subjective evaluation models** have been acknowledged covering different aspects of open data evaluation, namely, (i) maturity assessment, (ii) readiness assessment, (iii) post adoption and (iv) impact assessment. The latter group of evaluation models could be either qualitative (in their first stages) or quantitative (more advanced ones). Finally, some **objective, obsolete and quantitative indexes** are presented within this section.

8.2.1 Subjective Evaluation Models

This section emphases on the subjective models of evaluation in the domain of information systems. The above research streams of information systems evaluation are concentrated in capturing users' opinions about different aspects (perceived ease of use, perceived usefulness, attitude, intention to use, future behaviour etc.) of the system under evaluation. They formally raise questions and quantifies them using a five or seven-point Likert-scale towards the measurement of users' judgements.

8.2.1.1 IS Evaluation

Extensive research has been conducted on IS evaluation in the last 20 years (Farbey et al., 1999; Gunasekaran, Ngai, & McGaughey, 2006; Irani & Love, 2008; Smithson & Hirscheim, 1998; Willcocks & Graeser, 2001). Its main conclusion has been that IS evaluation is a difficult and complex task, since IS offer various types of benefits, both financial and non-financial, and also tangible and intangible ones, which differ among the different types of IS. Therefore, each particular type of IS requires a different evaluation methodology, which takes into account its particular objectives and capabilities. Smithson and Hirscheim (1998) distinguish between two basic directions of IS evaluation.

The first one is 'efficiency-oriented', evaluating IS performance with respect to some predefined technical and functional specifications; it focuses on answering the question of whether the IS 'is doing things right'. The second direction is 'effectiveness-oriented', evaluating to what extent the IS supports the execution of business-level tasks or the achievement of business-level objectives; it focuses on answering the question of whether the IS 'is doing the right things'. The conclusions of this research stream indicate that a comprehensive methodology for evaluating a particular type of IS should include evaluation of both its efficiency and its effectiveness, based on its particular objectives and capabilities.

8.2.1.2 Technology Acceptance Models

Another central topic in IS research has been the identification of characteristics and factors of IS that affect the intention to use them and finally the extent of its actual usage. This research has led to the development and extensive validation of the Technology Acceptance Model (TAM) and its subsequent extensions (Davis, 1989; Schepers & Wetzels, 2007; Venkatesh & Davis, 2000; Venkatesh, Morris, Davis, & Davis, 2003; Wixom & Todd, 2005). According to this model two characteristics of an IS, its perceived usefulness (= the degree to which users believe that using it will enhance their job performance) and its perceived ease of use (=the degree to which users believe that using it would require minimal effort), are the main determinants of individuals' intention to use it in the future and finally the actual use of it. The conclusions of this IS acceptance research stream indicate that a methodology for evaluating a particular type of IS should assess its ease of use, usefulness and users' intention to use it in the future.

Technology Acceptance Models have been influenced by Theory of Reasoned Action introduced by Fishbein & Ajzen, in 1975, and Theory of Planned Behavior (TPB) introduced by Ajzen, in 1991 and "posits that perceived usefulness and perceived ease of use determine an individual's intention to use a system with intention to use serving as a mediator of actual system use". Perceived usefulness is also seen as being directly impacted by perceived ease of use. Researchers have simplified TAM by removing the attitude construct found in TRA from the current specification by Venkatesh and Davis, in 2000, and Venkatesh et al. (2003). Attempts to extend TAM have generally taken one of three approaches:

(a) by introducing factors from related models,
(b) by introducing additional or alternative belief factors, and
(c) by examining antecedents and moderators of perceived usefulness and perceived ease of use as concluded by Wixom and Todd, in 2005.

TRA and TAM, both of which have strong behavioural elements, assume that when someone forms an intention to act, that they will be free to act without limitation. In practice constraints such as limited ability, time, environmental or organizational limits, and unconscious habits will limit the freedom to act is an information systems theory that models how users accept and use a technology. The model suggests that when users are presented with a new technology, a number of factors influence their decision about using it, but the two main factors (according to Davis et al., 1989):

- Perceived usefulness (PU), defined by F. Davis as "the degree to which a person believes that using a particular system would enhance his or her job performance".
- Perceived ease-of-use (PEOU) – defined by F. Davis as "the degree to which, a person believes that using a particular system would be free from effort"(Fig. 8.1).

Each of these two factors can be developed into a detailed set of variables for each particular type of Information System. Based on this framework, extensive research has been conducted for understanding better, and predicting user acceptance of various types of Information Systems (as concluded by Schepers & Wetzels,

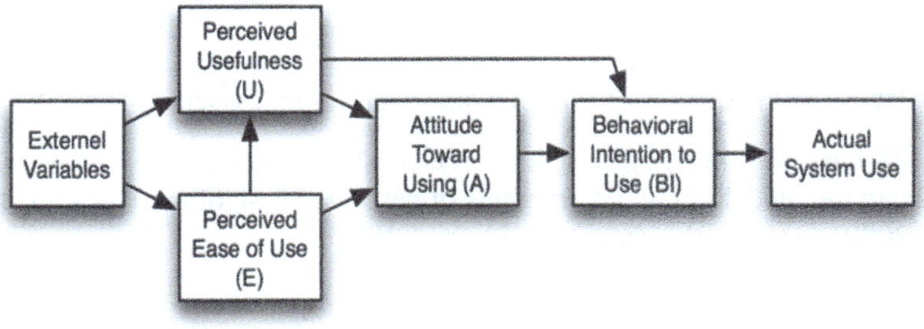

Fig. 8.1 Technology acceptance model

2007). As referred by Venkatesh and Davis (2000, TAM is continued to expand, the two major upgrade being the TAM2 and the Unified Theory of Acceptance and Use of TAM2 explains perceived usefulness and usage intentions in terms of social influence and cognitive instrumental processes. Both social influence processes (subjective norm, voluntariness, and image) and cognitive instrumental processes (job relevance, output quality, result demonstrability, and perceived ease of use) significantly influenced user acceptance.

In articles by Venkatesh et al. (2003), and Venkatesh and Zhang (2010) it is being shown that the theory of acceptance and use of technology (UTAUT) is useful to enrich one's understanding of research on technology adoption. The theory was developed through a review and consolidation of the constructs of eight models that earlier research had employed to explain information systems usage behaviour. The theory uses constructs of: theory of reasoned action, technology acceptance model, motivational model, theory of planned behaviour, a combined theory of planned behaviour/technology acceptance model, model of PC utilization, innovation diffusion theory, and social cognitive theory. UTAUT provides the rationale for the survey questions.

According to Venkatesh, UTAUT identifies

1. 3 direct determinants of behavioural intention to use a technology:
   (a) Performance expectancy (PE): *the degree to which an individual believes that using the system will help him or her to attain gains in job performance*
   (b) Effort expectancy (EE): *the degree of ease associated with the use of the system*
   (c) Social influence (SI): the degree to which an individual perceives that important others believe he or she should use the new system
2. 2 direct determinants of technology use
   (a) Behavioural intention
   (b) Facilitating conditions (FC): the degree to which an individual believes that an organizational and technical infrastructure exists to support use of the system
3. 4 contingencies
   (a) CG-1: Gender
   (b) CG-2: Age
   (c) CG-3: Experience with the technology
   (d) CG-4: Voluntariness of use (mandatory or voluntary setting) (Fig. 8.2)

TAM3 have also been proposed by Venkatesh and Bala, 2008. They combine TAM2 and the model of the determinants of perceived ease of use (by Venkatesh & Davis, 2000) to end to the above extended model.

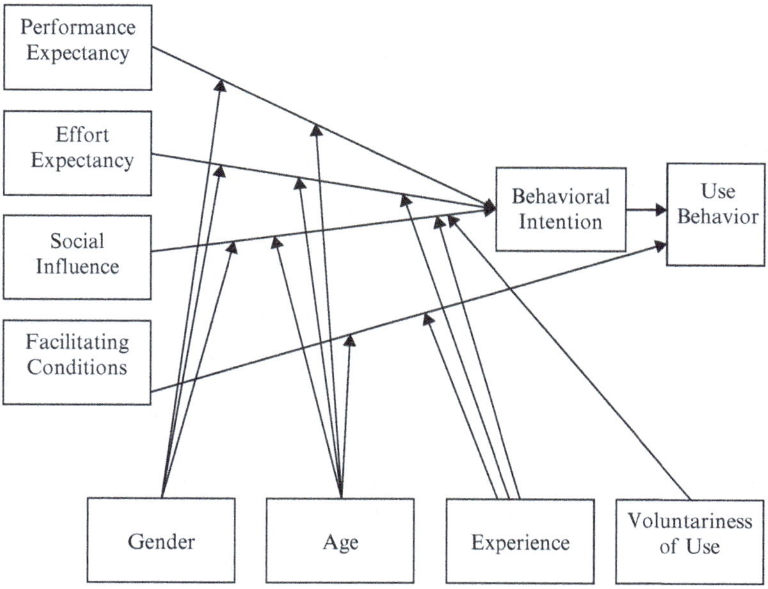

Fig. 8.2 Unified theory of acceptance and use of technology

8.2.1.3 Information Systems Success Models

Another research stream that can provide useful elements is the IS success research (DeLone & McLean, 1992, 2003; Seddon, 1997). The most widely used IS success model has been developed by DeLone and McLean (1992). It proposes seven IS success measures, which are structured in three layers: 'information quality', 'system quality' and 'service quality' (at the first layer), which affect 'user satisfaction' and also the 'actual use' of the IS (at the second level); these two variables determine the 'individual impact' and the 'organizational impact' of the IS. Seddon (1997) proposed a re-specification and extension of this model, which includes perceived usefulness instead of actual use. The conclusions of this research stream indicate that IS evaluation should adopt a layered approach based on the above interrelated IS success measures (information quality, system quality, service quality, user satisfaction, actual use, perceived usefulness, individual impact and organizational impact) and also on the relations among them.

The IS success theoretical model, was first developed by William H. DeLone and Ephraim R. McLean in 1992. The most widely used System Success Model is the one by DeLone and McLean: Model of IS success, developed in 2003. It proposes seven IS success measures, which are structured in three layers:

1. First layer: 'information quality', 'system quality' and 'service quality'
2. Second layer: Affecting 'user satisfaction' and
3. Third layer: 'actual use' of the IS.

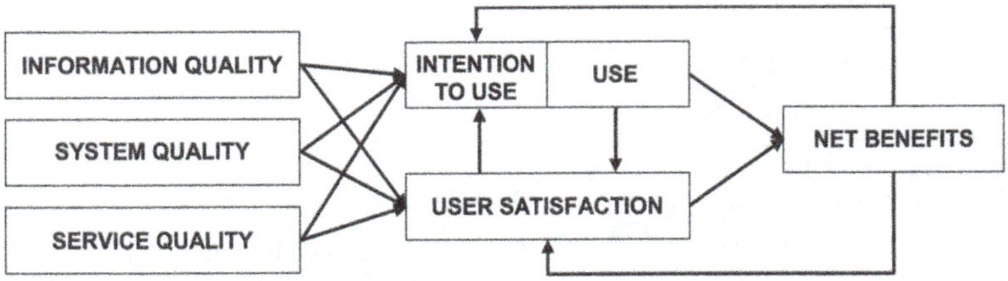

Fig. 8.3 DeLone and McLean: model of IS success. (Source: DeLone and McLean (2003))

Finally, these two variables determine the 'individual impact' and the 'organizational impact' of the IS. Seddon, in 1997, proposed a re-specification and extension of this model, which includes perceived usefulness instead of actual use. From this research stream, it has been concluded that IS evaluation should adopt a layered approach based on the above interrelated IS success measures (information quality, system quality, service quality, user satisfaction, actual use, perceived usefulness, individual impact and organizational impact) and on the relations among them (Fig. 8.3)

8.2.1.4 E-services Evaluation

The emergence of numerous Internet-based e-services (e.g. information portals, e-commerce, e-banking, e-government, etc.) lead to the development of specialised frameworks for evaluating them (Fassnacht and Koese, 2006; Lu and Zhang, 2003; Rowley, 2006; Saha and Grover, 2011; Sumak, Polancic, & Hericko, 2009); extensive reviews of this research are provided from Rowley (2006) and Sumak et al. (2009). These frameworks suggest useful e-services evaluation dimensions and measures. Most of them assess the quality of the capabilities that the e-service provides to its users (being oriented towards the abovementioned efficiency evaluation). Some others assess the support it provides to users for performing various tasks and achieving various objectives (being oriented towards the above-mentioned efficiency evaluation).

SERVQUAL is a service quality framework. SERVQUAL was developed in the mid-eighties by Parasuraman et al. 1998. and was initially used in a marketing context. Later Zeithaml (2002) applied to IS as a measure of success. SERVQUAL model consists of 22 service quality measures that are organized in five dimensions:

- tangibles (appearance of physical facilities, equipment, personnel and communication materials)
- reliability (ability to perform the promised service dependable and accurately)
- responsiveness (willingness to help customers and provide prompt service)
- assurance (knowledge and courtesy of employees and ability to convey trust and confidence)
- empathy (provision of caring, individualized attention to customers)

Parasuraman, Zeithaml, and Malhotra (2005) extended SERVQUAL for the evaluation service quality in web-based environments. So they named E-S-Qual, e-service quality. E-S-QUAL Scale, consisting of 22 items on four dimensions:

- Efficiency: The ease and speed of accessing and using the site.
- Fulfilment: The extent to which the site's promises about order delivery and item availability are fulfilled.
- System availability: The correct technical functioning of the site.
- Privacy: The degree to which the site is safe and protects customer information.

Parasuraman also tries to measure the quality of recovery service provided by Web sites. The e-recovery service quality scale (E-RecSQUAL) consisting of 11 items on three dimensions:

- Responsiveness: Effective handling of problems and returns through the site.
- Compensation: The degree to which the site compensates customers for problems.
- Contact: The availability of assistance through telephone or online representatives.

However, most of the above frameworks do not include advanced ways of processing the evaluation data collected from the users, in order to maximize the extraction of value-related knowledge from them. They include mainly simple calculations of average values of all evaluation measures and dimensions; the relations among the proposed evaluation dimensions and measures, which could form the basis for advanced multi-dimensional statistical analysis, are not exploited all for drawing more insights. Section 8.3 presents an evaluation framework based on value models prioritising future developments (Charalabidis, Loukis, & Alexopoulos, 2014).

8.2.1.5 Maturity Models

In the open data domain, maturity is defined as a measurement of the ability of an organization or a country for continuous improvement. The higher the maturity, the higher the probability of transforming incidents into improvement either in their quality or in their use. Most of the maturity models are subjective in terms of model conceptualisation and qualitative, but the more advanced ones specify quantitative techniques towards the assessment of their maturity and proposition of the next steps of development (Solar, Daniels, López, & Meijueiro, 2014). Concerning open data maturity models several authors have presented different stages to assess and diagnose open data (Alexopoulos, 2016; Kalampokis, Tambouris, and Tarabanis, 2011a; Reggy, 2011).

Open Government Data is a sub-domain of e-government and as such it follows its general principals. The overall approach to maturity in e-government has so far been evolutionary as stated by Krishnan, Teo, & Lim, in 2013 – governments are believed to progress through certain stages. Stages of growth models, in general, receive criticism for their limited applicability and misleading normative values: in practice, several stages may occur simultaneously. Furthermore, the models are constructed in such a way that preceding stages appear to be "worse" than subsequent ones as demonstrated by K. V. Andersen & Henriksen, in 2006. The contemporary debate about e-government maturity has shifted from supply-side models to user-centric maturity indicators.

The view of e-government maturity as a function of integration and organizational and technological complexity in the early model by Layne and Lee (2001) can be considered a manifestation of technology bias. An alternative vision is proposed in the model by K. N. Andersen, Medaglia, and Henriksen (2012), which uses citizen orientation and activity centricity as the primary criteria for deriving the four e-government maturity stages, namely, cultivation, extension, maturity, and revolution (Susha, Zuiderwijk, Janssen, & Gronlund, 2014).

The recent study on the European data portal from Capgemini (Carrara, Chan, Fischer, & Steenbergen, 2015) has developed a maturity model for the EU28 countries regarding their portals development. "*To provide an accurate estimate of the benefits of Open Data, one first needs to look at the Open Data Maturity per country and how this maturity has evolved.*" There are substantial differences between the EU28+ countries when measuring the progress made so far in terms of Open Data. To take these discrepancies into account count, a model was developed to classify the maturity of a country with regards to Open Data. Based on the scores on several indicators, countries were compared in terms of their maturity. This resulted in a matrix with different scores per country. A country can be classified as being either a Trend Setter, Follower, Advanced Beginner or Beginner. The model showed that in 2005, 63% of the Member States could be classified as a Beginner whilst not a single country could be classified as a Trend Setter. These numbers changed substantially over the past 10 years. In 2015, 31% of the countries can be classified as a Trend Setter whereas only 19% is still a Beginner. By 2020 all countries will have a fully operating portal. Additionally, countries will also introduce improvements to increase their Open Data Maturity.

8.2.1.6 Readiness Assessment

Opening up data by public bodies is a complex and ill-understood activity. Although many public bodies might be willing to open up their data, they lack any systematic guidance. A readiness assessment framework aims at the determination: (a) of the status of an organisation to open up its data for re-use as well as (b) of the data status in terms of format, licencing, means of provision in order to be useful for re-use. It is dealing with organisational issues covered in Chap. 4 and includes the processes of the open data life cycle towards publication of data covered in Chap. 2. It could also be referring to issues on deciding whether to open data or to publish them in restricted access. They might provide solutions dealing with privacy-sensitive data, deletion policies, publishing after embargo periods instead of not publishing at all. Examples of readiness assessment frameworks have been proposed by Zuiderwijk et al. (2012c) and the World Bank (2013a) through the creation of the Open Data Readiness Assessment tool.

The process of opening up public sector data demands considerable changes in the public sector, such as changes in the funding and reward systems of organizations. However, it is usually not possible to explain how those types of e-Government initiatives evolve over a certain period of time by the current e-Government linear progression models and the development of composite e-Government services is usually ad-hoc. The questions that are expected to easily rule out opening up a certain dataset are placed on top of the list, whereas questions that require further examination are placed at the bottom of the list. This is done so that data that cannot be opened are quickly identified. Aspects of institutional theory were taken into account by considering the risk avoiding governmental culture. For instance, due to the fear of wrongful interpretations of the data and the impact of wrongful interpretation on the organization, such as hitting the news with a damaged reputation, guidance is provided to make the chance on wrongful interpretations as small as possible.

8.2.1.7 Post Adoption

We define post-adoption stage what Hazen, Overstreet, and Cegielski (2012) drew from numerous literature where they tried to uncover whether the ambiguity after the innovation or technology has been accepted in an organization. The final stage of post adoption assessment is called "incorporated". This incorporated stage may include three post-adoption activities where it includes acceptance, routinization, and assimilation (Nurakmal & Hamid, 2012). Several studies have proven that post-adoption assessment frameworks are useful in the investigation of a wide range of IT innovations in an organization.

Although, some studies have found new factors or measures to influence technology adoption, the factors will still fall in either one of the three already identified constructs. This shows that the three antecedents (technology, organization, environment) are dynamic and can be manipulated with various factors that influence organization to adopt innovation or technology. In (Nurakmal & Hamid, 2012), Tornatzky antecedents where further extended to the stages of post-adoption described by Hazen et al. (in 2012), which consist of assimilation, routinization, and acceptance stage. The actual factors in technology, organization and environment context will were mapped with the data gathered. Each of Tornatzky antecedents was assumed to have influence on post- adoption stages. Therefore, a set of hypotheses can be construct to test the relationship.

8.2.1.8 Impact Assessment

The impact of opening up data is often debated and espoused as the primary reason for publishing Open Data. While recourse to its economic and democratic impact is seen as a useful driver for publicizing more data, it is rarely easy to quantify the impact this initiative has on business and society. So far, efforts at measuring impact have been mixed and unable to produce concrete results on the usefulness of Open

Data. The crux of the issues lies in the fact that merely opening up datasets does not automatically mean that the public can use them meaningfully or that business can profitably utilize them.

Publication is a prerequisite, but also public interest and regular recourse to information is needed to ensure that large benefits are reaped. Apart from access, the impact of open data depends crucially on engagement, ability to analyse, and draw conclusions from information, and a suitable institutional and economic environment that is receptive of such innovation. In fact, barriers to usage of open data are sometimes seen as so high that some authors argue that open data empowers the already empowered – the highly educated persons and sophisticated businesses that can extract value from public information. All this is likely to put real-world open data impact in perspective, as it is likely smaller and more unequal than usually discussed in public policy circles.

Impact measurement has tended to center around two large groups of metrics – quality, usage, and access on the one hand; and results-based metrics on the other (Gerunov, 2016). As demonstrated in (Gerunov, 2016), impact metrics need to quantify both economic and political benefits brought about by the totality of open data, and also take account of the distribution of those benefits. We can outline three major approaches to measuring this impact depending on the level on which measurement takes place:

1. In macro-level approaches the researchers assume that opening data should have an overall effect on the economy and society, and therefore measurement and assessment should take place at the aggregate level. Since OGD is supposed to stimulate information and improve the public environment, it should be the case that it is associated with a measure of technological development such as total factor productivity (TFP).
2. Meso-level approaches look at the impact of OGD at the sector to which it pertains. Opening data in a specific sector should bring notable improvement in it, which can be seen in some predetermined data indicators. For example, opening procurement data should lead to more transparency and less corruption and thus lower the price for reference orders.
3. Micro-level approaches focus on specific datasets or groups of datasets, and follow them through their lifecycle. By doing this, the researcher gets a full and nuanced picture of usage, impact, and benefit distribution. The most common micro-level approach is the case study whereby each OGD dataset usage is described in detail, giving the context and measuring benefits to different stakeholders. Case studies generally use a mixed method design and serve as an excellent illustration of OGD potential. They can thus be leveraged as a powerful argument in favor of openness. The main issues with this approach are that it fails to scale well and is suffering from observer bias. What is more, this method poses challenge to the researcher to exhaustively identify all the benefits of the dataset and to quantify the full set of externalities. This is counterbalanced by the fact that the analysis is more intuitive to make and ends in tractable results. The method of choice for measuring impact naturally differs across situations and

has to adapt to the context of specific data openness. What is most important is not to overlook this key aspect of OGD policy.

The recent study on the European data portal from Capgemini (Carrara, Chan, et al., 2015) has collected, assessed and aggregated economic evidence to forecast the benefits of the re-use of Open Data for the EU28+. This study falls into the first two categories of impact assessment. The expected impact of the Open Data policies and the development of data portals is to drive economic benefits and further transparency. Four key indicators are measured: direct market size, number of jobs created, cost savings, and efficiency gains. Between 2016 and 2020, the market size of Open Data is expected to increase by 36.9%, to a value of 75.7 bn EUR in 2020. The forecasted public sector cost savings for the EU28+ in 2020 are 1.7 bn EUR. Efficiency gains are measured in a qualitative approach. A combination of insights around efficiency gains of Open Data, and real-life examples is provided.

8.2.2 *Objective Evaluation Models*

Since the publication of the eight principles of open government data, and the "five stars" test proposed by Bizer, et al. (2011), several authors and institutes have presented different objective criteria to assess and diagnose Open Data based on the development of quantitative indexes, such as the Open Data Institute,[1] the Open Data Research Network,[2] the Open Knowledge Foundation,[3] the Open Data 500,[4] the Open Data Monitor,[5] the Dynamic Linked Data Observatory,[6] the Open Data Barometer[7] and others. These indexes utilise specific metrics for the measurement of different aspects (e.g. data quality, popularity, and user feedback).

For instance, metrics such as number of views, downloads and reuses could be used to measure the popularity of open datasets. Metrics such as (a) accuracy: defined by the number of accurate values divided by the total number of all values, (b) completeness: number of non-null values divided by the total number of all values and (c) timeliness: number of values that are up-to-date divided by the total number of values formulate the quality index of a dataset. Another objective and quantitative evaluation model has been developed for the evaluation of linked data quality by Kontokostas, Westphal, Auer, Hellmann, et al. (2014b).

---

[1] https://theodi.org/

[2] http://www.opendataresearch.org/

[3] https://okfn.org/

[4] http://www.opendata500.com/

[5] http://opendatamonitor.eu/frontend/web/index.php?r=dashboard%2Findex

[6] http://swse.deri.org/dyldo/

[7] http://opendatabarometer.org/

8.3 Applying Evaluation Models on Open Data

This section presents different examples of different applications of Open Data assessment based on the analysed models in Sect. 8.2. The presented models have been adapted to the assessment of open data and their platforms assessing various aspects of open data using both objective and subjective methods of evaluation.

8.3.1 Adapting IS Success Model on Open Data Evaluation

The model proposed by Charalabidis et al. (2014), for the evaluation of the advanced second generation of OGD, was primarily based on the IS success model (adopting a layered evaluation approach, and including measures of both information and system quality, and also of user satisfaction and individual impact). The model aims at predicting the future behaviour of its users. It is a subjective model based on user opinions collected with the form of a questionnaire.

Particularly value dimensions are organized in three value layers adopting the structure proposed by (Loukis et al., 2012; Pazalos et al., 2012), which correspond to efficiency (value associated with the capabilities it offers to the users), effectiveness (value associated with the support of users for achieving their user-level and provider-level objectives) and future behavior (value associated with users' future behavior) respectively.

The first efficiency layer includes eight value dimensions in total. Three of them concern the user-level capabilities offered by the OGD infrastructure: data provision capabilities data search and download capabilities and user-level feedback capabilities. These value dimensions are expected to affect the 'support for achieving user-level objectives' value dimension of the second. The next three value dimensions of the first layer are: performance, accessibility and data processing capabilities. They are expected to affect both the 'support for achieving user-level objectives' and the 'support for achieving provider-level objectives' value dimensions of the second layer. The final two dimensions of the first layer concern the provider-level capabilities offered by the OGD infrastructure: data upload capabilities and provider-level feedback capabilities. They are expected to affect the 'support for achieving provider-level objectives' value dimension of the second layer. The second effectiveness layer includes the abovementioned two value dimensions concerning the support provided by the OGD infrastructure for achieving user-level and provider-level objectives respectively. Lastly, the third layer includes one value dimension associated with users' future behavior.

The above 11 value dimensions were further elaborated, and for each of them a number of individual value measures were defined. Each of these value measures was then converted to a question to be included in a questionnaire to be distributed to users of the infrastructure (who act both as data users and providers). The Table 8.1 presents the measures for each dimension:

Table 8.1 Value models – dimension measures

Data Provision Capabilities (DPV)	
DPV1	The platform provides a large number of datasets
DPV2	The platform provides datasets useful to me
DPV3	The platform provides to me complete data with all required fields and detail
DPR4	The platform provides accurate and reliable data on which I can rely for my studies
DPV5	There are datasets from many different thematic areas (economy, health, education, etc.)
DPV6	There are datasets from many different countries
DPV7	The platform provides sufficiently recent data
Data Search and Download Capabilities (DSD)	
DSD1	The platform provides strong dataset search capabilities using different criteria.
DSD2	The platform provides several different categorizations of the available datasets, which assists significantly in finding the datasets I need.
DSD3	The platform enabled me to download datasets easily and efficiently.
DSD4	The datasets are in appropriate file/data formats that I can easily use.
DSD5	The datasets have also appropriate and sufficient metadata, which allowed me to understand these data and also how and for what purpose they were collected.
DSD6	The platform provides strong API for searching and downloading datasets (data and metadata)
User-level Feedback Capabilities (UFB)	
UFB1	The platform provides good capabilities for giving feedback on the datasets I download, e.g. for rating datasets, for entering textual comments on them.
UFB2	The platform provides good capabilities for reading available feedback of other users of datasets I am interested in, e.g. ratings, comments.
Ease of Use (EOU)	
EOU1	The platform provides a user friendly and easy to use environment.
EOU2	It was easy to learn how to use the platform.
EOU3	The web pages look attractive.
EOU4	It is easy to perform the tasks I want in a small number of steps.
EOU5	The platform allows me to work in my own language.
EOU6	The platform supports user account creation in order to personalize views and information shown
EOU7	The platform provides high quality of documentation and online help.
Performance (PER)	
PER1	The platform is always up and available without any interruptions.
PER2	Services and pages are loaded quickly.
PER3	I did not realize any bugs while using the platform.
Data Processing Capabilities (DPR)	
DPR1	The platform provides good capabilities for data enrichment (i.e. adding new elements – fields)
DPR2	The platform provides good capabilities for data cleansing (i.e. detecting and correcting ubiquities in a dataset)
DPR3	The platform provides good capabilities for linking datasets.
DPR4	The platform provides good capabilities for visualization of datasets

(continued)

Table 8.1 (continued)

Data Upload Capabilities (DUP)	
DUP1	The platform enabled me to upload datasets easily and efficiently.
DUP2	The platform enabled me to prepare and add the metadata for the datasets I uploaded easily and efficiently.
DUP3	The platform provides good capabilities for the automated creation of metadata.
DUP4	The platform provides good capabilities for converting datasets' initial metadata in the metadata model of the platform easily and efficiently.
DUP5	The platform provides strong API for uploading datasets (data and metadata)
Provider-level Feedback Capabilities (PFB)	
PFB1	The platform allows me to collect user ratings and comments on the datasets I publish.
Support for Achieving User-level Objectives (SUO)	
SUO1	I think that using this platform enables me to do better research/inquiry and accomplish it more quickly
SUO2	This platform allows drawing interesting conclusions on past government activity
SUO3	This platform allows creating successful added-value electronic services
Support for Achieving Provider-level Objectives (SPO)	
SPO1	The platform enables opening and widely publishing datasets with low effort and cost.
Future Behaviour (FBE)	
FBE1	I would like to use this platform again.
FBE2	I'll recommend this platform colleagues.

According to model (Charalabidis et al., 2014) the above value can be adapted based on the capabilities offered by the particular second generation OGD infrastructure under evaluation (e.g. additional value dimensions can be added corresponding to additional capabilities it might offer). Furthermore, the above approach can be used for the evaluation of first generation OGD infrastructures as well, which are characterized by clear distinction between data providers and data users, by defining and estimating one value model for the former and one value model for the latter (Fig. 8.4).

8.3.2 Adapting UTAUT on Open Data Evaluation

According to (Zuiderwijk, Janssen, & Dwivedi, 2015) the ability to use open data partly depends on the availability of open data technologies. Therefore, the acceptance and use of Information Technology has been of significant importance for Information Systems research and practice. The UTAUT is an often used model that examines Information Technology acceptance and use.

Thus, a subjective model developed by (Zuiderwijk et al., 2015) to obtain the acceptance and use of open public sector from actual users of these data. The model has the form of questionnaire and is designed following the construct of the UTAUT research model with a modification. At the table below are seen the questions which were asked. Some of the questions are answered with the a five-point Likert scale to

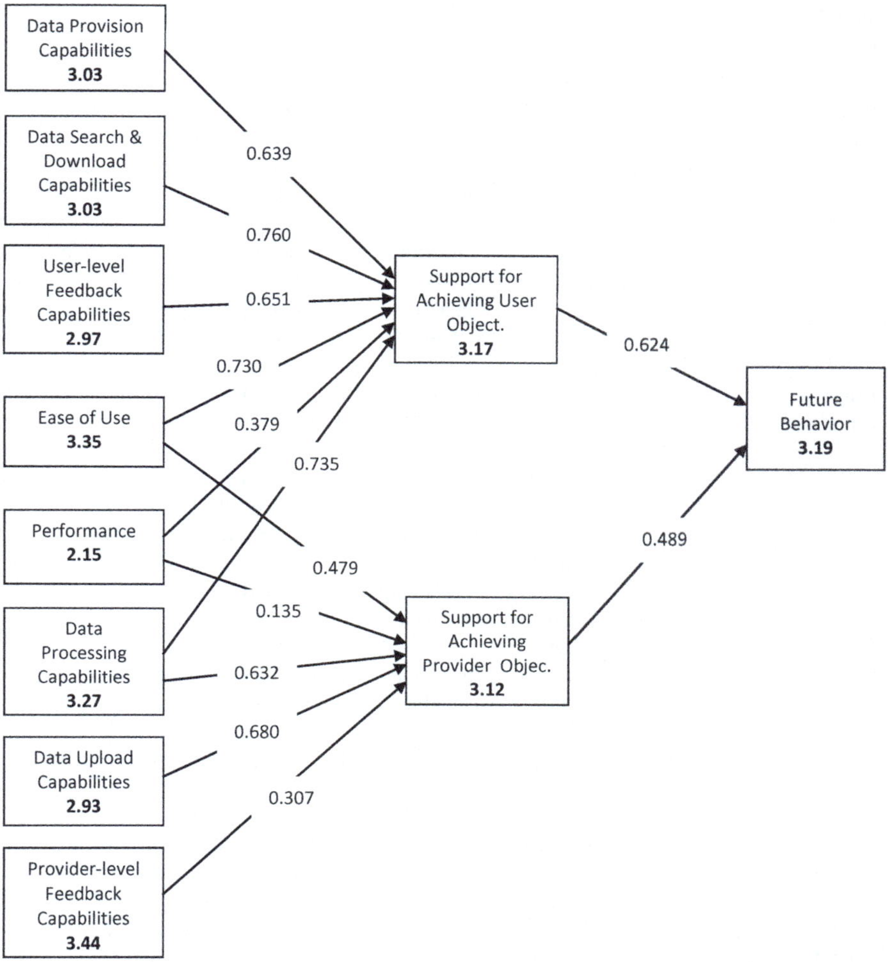

Fig. 8.4 Value model for Advanced Open Data Platforms Evaluation

which extent they agreed with the statement, ranging from "strongly disagree" to "strongly agree (Table 8.2).

8.3.3 Creation of an Objective Model for Open Data Platforms Assessment

Another approach analyses the main characteristics of OGD data portals from different perspectives and implemented by (Alexopoulos, Loukis, Petychakis, & Charalabidis, 2015). The model has focused on the objective evaluation of Open Data sources characteristics and it was applied for the assessment of the Greek open

Table 8.2 Questionnaire for the UTAUT model

UTAUT construct	Questionnaire item (statement or question)	Type of outcome
Performance expectancy (PE)	Using open public sector data is of benefit tome (PE1)	Five-point Likert scale (strongly disagree-strongly agree)
	Using open public sector data will enable me to accomplish my research more quickly (PE2)	Five-point Likert scale (strongly disagree-strongly agree)
	Using open public sector data will increase my productivity (PE3)	Five-point LIkert scale (strongly disagree-strongly agree)
	Using open public sector data improves my performance in my job (PE4)	Five-point Likert scale (strongly disagree-strongly agree)
Effort expectancy (EE)	It will be easy for me to become skillful at using open public sector data (EE1)	Five-point Likert scale (strongly disagree-strongly agree)
	Learning to use open public sector data will be easy for me (EE2)	Five-point Likert scale (strongly disagree-strongly agree)
	I clearly understand how to use open public sector data (EE3)	Five-point Likert scale (strongly disagree-strongly agree)
	I do not have difficulty in explaining why using open public sector data may be beneficial (EE4)	Five-point Likert scale (strongly disagree-strongly agree)
Social influence (SI)	People who influence my behavior think that I should use open public sector data (SI1)	Five-point Likert scale (strongly disagree-strongly agree)
	People who are important to me (e.g. family, friends) think that I should use open public sector data (SI2)	Five-point Likert scale (strongly disagree-strongly agree)
	People who are important to me (.e.g. colleagues) think that 1 .should use open public sector data (SI3)	Five-point Likert scale (strongly disagree-strongly agree)
Facilitating conditions (FC)	I have the resources necessary to use open public sector data(FC1)	Five-point Likert scale (strongly disagree-strongly agree)
	Open public sector data is compatible with other systems that I use (FC2)	Five-point Likert scale (strongly disagree-strongly agree)
	A specific person or group is available for assistance with di faculties concerning the use of open public sector data (FC3)	Five-point Likert scale (strongly disagree-strongly agree)
Behavioral· intention (BI)	I intend to use open public sector data in the future (BI1)	Five-point Likert scale (strongly disagree-strongly agree)
	I predict that I will use open public sector data in the future (BI2)	Five-point Likert scale (strongly disagree-strongly agree)
	I plan to use open public sector data in the future (BI3)	Five-point Likert scale (strongly disagree-strongly agree)

(continued)

Table 8.2 (continued)

UTAUT construct	Questionnaire item (statement or question)	Type of outcome
Voluntariness of use (VU)	Although it might be helpful, using open public sector data is certainly not compulsory for my research or other activities (VU1)	Five-point Likert scale (strongly disagree-strongly agree)
	My research and other activities do not require me to use open public sector data (VU2)	Five-point Likert scale (strongly disagree-strongly agree)
	My superiors expect me to use open public sector data (VU3) (R)	Five-point Likert scale (strongly disagree-strongly agree)
	My use of open public sector data is voluntary (it is not requited by my superiors/research/other activities) (VU4)	Five-point Likert scale (strongly disagree-strongly agree)
Gender (G)	Are you male or female? (G)	Multiple choice (male or female)
Age (A)	What is your age? (A)	Eight-point scale (under 18–61 or over)
Purpose of use (P)	To what extent are the following purposes important for your use of open public sector data? (P)	Five-point Likert scale (very unimportant-very important)
Type of data (T)	Which of the following types of open data from the public sector do you use or have you used? (T)	Multiple choice (type of public sector data: geographic, legal, meteorological, social, transport, business, other, namely...)

Each statement or question was given a code, referring to the UTAUT construct. The items labeled "(R)" are reverse-coded

data sources. Four dimensions/perspectives have been defined evaluating different aspects of the sources offering open data. These perspectives are as follows:

1. Thematic Analysis Perspective: It includes analysis of the thematic categories of the datasets provided by the OGD sources.
2. Functional Analysis Perspective: It includes analysis of the functionalities provided by the OGD sources.(Datasets discovery, Data provision, Language, Visualizations and feedback)
3. Semantic Analysis Perspective: It includes analysis of the use of Semantic Web technologies for the representation and structure of OGD. using the well established 5-stars Berner Lee's rating system for open data and then an analysis of the metadata and of licence information.
4. Technological Analysis Perspective: It includes analysis of the technologies and products that have been used for the development of the OGD source at the main technological layers: web server, Content Management System (CMS) or platform, user interface, data format and API.

8.3.4 Developing Maturity Models for Open Data

The maturity model concept stands for a model categorising the capabilities of OGD infrastructures through time as described in (Alexopoulos, Diamantopoulou, & Charalabidis, 2017). OGD portals are distinguished in two main categories: traditional and advanced infrastructures. The identified elements of OGD portals are categorized in 4 dimensions as it is seen above: general; information quality; system quality and service quality. Last three dimensions are based in IS Success model. Each of these elements defined by specific values. Thus, this maturity model constitutes an objective assessment. According to Alexopoulos the developed maturity model will guide policy makers by firstly identify the current level of their organization and secondly design an efficient implementation to the required state (Table 8.3).

Another more advanced maturity model has been created by (Solar, Concha, & Meijueiro, 2012). The proposed maturity model, named OD-MM (Open Data Maturity Model) assesses the commitment and capabilities of public agencies in pursuing the principles and practices of open data. It is a subjective (users' opinions) and quantitative model which consists of a three level hierarchical structure, called domains, sub-domains and critical variables. Four capacity levels are defined for each of the 33 critical variables distributed in nine sub-domains in order to determine the organization maturity level. The model is a very valuable diagnosis tool for public services, given it shows all weaknesses and the way (a roadmap) to progress in the implementation of open data.

8.3.5 Institutional Readiness Assessment for Open Data Publishers

The framework developed by (Agbabiaka & Ojo, 2014) for assessing institutional readiness into four main areas: people readiness; system readiness; technology readiness and process readiness. The framework focused on system readiness that constitutes in various sub-dimensions for assessment based on subjective evaluation as described below. Each of sub-dimensions can be assessed with the following values: no progress, some progress, real progress is being made, ready and effective corresponding to the following readiness level: poor, low, medium and high (Table 8.4).

8.4 Metrics Classification

The taxonomy of open data evaluation metrics is based on the "information system success" model, we are going to categorize different evaluation measures and benchmarks for the evaluation of data (Information Quality), platforms offering them (System Quality) and additional capabilities of those systems (Service Quality). Figure 8.5 presents an overview of the main classification categories.

Table 8.3 Maturity model for OGD portals

		Traditional OGD infrastructures		Advanced OGD infrastructures	
	Time	Point zero	**1st generation**	**2nd generation**	**3rd generation**
General	**Internet presence**	OGD existence in silos accessed by application	OGD web presence	OGD web presence	OGD web presence
	Users	Distinction between data providers and data users	Distinction between data providers and data users	Data pro-sumers	Data pro-sumers
	Open government level	Initial: Information broadcasting	Data transparency: Processes and performance	Open participation: Data quality, public feedback, conversation, voting, interactive communications, crowd-sourcing	Open collaboration: Interagency and with the public, co-creating value-added services
	Value	N/A	Transparency & accountability	Participation	Efficiency & innovation
Information quality	**Thematic perspective**	N/A	Statistical, economical, census	Law, transportation, GIS	All categories with proper data modelling
	Format	.xls, .pdf	html, .xls, .pdf	+ .csv + URLs	+ Linked data
	Metadata	Metadata ignorance or closed flat metadata	Metadata ignorance or closed flat metadata	Open metadata for humans or open reusable metadata + contextual or detailed metadata models	Linked open metadata 3-layer metadata model (flat, contextual, detailed)
	RDF-compliance	No	No	Partially yes	Yes
System quality	**Functionality**	N/A	Basic Web 1.0	Advanced Web 2.0	Supporting value creation
	Type	N/A	OGD direct provision portals	OGD direct provision & OGD aggregators	Collaboration spaces

Table 8.4 Framework for assessing institutional readiness

Measure	Definition
Governance readiness	This sub-dimension seeks to assess the presence of supporting mechanisms that will govern the process of preparing for the desired change
Legal & policy	The existence of relevant legal and policy framework that can aid or impede the desired change
Adaptive leadership	The availability of leaders within the organisation that can adapt, innovate and thrive in complex, challenging and uncertain environments
Resource readiness	Degree to which the resources of the agency can support the change. It assess whether the agency has effective financial policies and systems to support the viability and sustainability of the new change
Innovation capability	The degree to which the agency can create values from implementing new ideas and support the idea from conception to delivery.
Information sharing	Degree to which agency's policies, practices, legal framework support information sharing and willingness to embrace information sharing.
Collaboration & engagement	Degree to which the agency is willing to collaborate within itself and with other agencies as well as engage stakeholders and the public in the delivery of its services.
Open data readiness	Degree to which the agency is ready to make data available to other agencies and the public in a transparent way
Change management readiness	Degree to which the agency is prepared to adapt to the anticipated or desired change and evolve.
People readiness	The people factor is a critical component and perhaps the singular most important element of any organization's readiness to accept change. This section of the assessment will cover evaluation of leadership support readiness, the quality and competence level of staff, leadership development policy, etc.

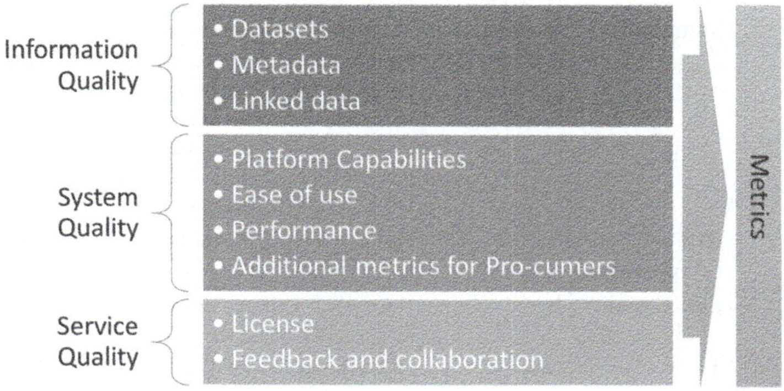

Fig. 8.5 Evaluation metrics classification

Additionally, different evaluation benchmarks for open data have been identified and categorised based on the following three aspects:

(i) The approaches and frameworks from previous relevant IS, concerning: IS evaluation (including in the methodology both efficiency and effectiveness measures), IS acceptance (including measures of ease of use, usefulness and future intentions), IS success (adopting a layered evaluation approach, and including measures of both information and system quality, and also of user satisfaction and individual impact) and e-services evaluation (including measures of both the quality of the capabilities offered to the users, and the support provided to them for achieving their OGD related objectives).

(ii) Potential users' requirements, which include data search, provision and download capabilities, data processing capabilities, data upload capabilities, and also users – providers feedback capabilities.

(iii) The high level technological aspects proposed in the methodologies for country and government agency level OGD initiatives' evaluation (such as data completeness, quality, quantity, format and metadata, search capabilities, users-providers communication capabilities, users' satisfaction, platform availability).

8.4.1 Information Quality

Information quality metrics are distinguished in three main dimensions: The datasets, the metadata and the linked data where relevant.

8.4.1.1 Data Sets

The dataset metrics are used to assess the data quality of the OGD. They examine the properties and the characteristics of the data (Table 8.5).

8.4.1.2 Metadata

Metadata: In addition to data quality, the second dimension examines the quality of the metadata including the necessary information for the description of the published data (Table 8.6).

8.4.1.3 Linked Data

The third aspect of information quality evaluation is the Linked Data where it is applicable. This dimension includes metrics to assess the quality of public data when they are linked (Table 8.7).

Table 8.5 Evaluation Metrics for Dataset

Dataset			
1	Uniqueness	Uniqueness is defined as the "degree to which data is free of redundancies, in breadth, depth and scope."	Behkamal, Kahani, Bagheri, and Jeremic (2014)
2	Primary	Data is as collected at the source, with the highest possible level of granularity, not in aggregate or modified forms.	https://public.resource.org/8_principles.html (2007)
3	Machine processable	Data is reasonably structured to allow automated processing	https://public.resource.org/8_principles.html (2007)
4	Non-discriminatory	Data is available to anyone, with no requirement of registration.	https://public.resource.org/8_principles.html (2007)
5	Non-proprietary	Data is available in a format over which no entity has exclusive control	https://public.resource.org/8_principles.html (2007)
6	Online and free	Information is not meaningfully public if it is not available on the internet at no charge, or at least no more than the marginal cost of reproduction. It should also be findable.	https://opengovdata.org/ (n.d.)
7	Permanent URI	Data should be made available at a stable internet location indefinitely and in a stable data format for as long as possible.	https://opengovdata.org/ (n.d.)
8	Safe to open	The Association of Computing Machinery's recommendation on open government (February 2009) stated, "government bodies publishing data online should always seek to publish using data formats that do not include executable content." executable content within documents poses a security risk to users of the data because the executable content may be malware (viruses, worms, etc.).	https://opengovdata.org/ (n.d.)
9	Designed with public input	The public is in the best position to determine what information technologies will be best suited for the applications the public intends to create for itself. Public input is therefore crucial to disseminating information in such a way that it has value.	https://opengovdata.org/ (n.d.)
10	Accessibility	Data is available to the widest range of users for the widest range of purposes. This information is easily retrievable. This information is easily accessible. This information is easily obtainable. This information is quickly accessible when needed.	Lee, Strong, Kahn, and Wang (2002)

(continued)

Table 8.5 (continued)

Dataset			
11	Appropriate amount	This information is of sufficient volume for our needs. The amount of information does not match our needs. The amount of information is not sufficient for our needs. The amount of information is neither too much nor too little.	Lee et al. (2002)
12	Completeness	All public data is made available. Public data is data that is not subject to valid privacy, security or privilege limitations. This information includes all necessary values. This information is incomplete. This information is complete. This information is sufficiently complete for our needs. This information covers the needs of our tasks. This information has sufficient breadth and depth for our task.	Lee et al. (2002)
13	Concise representation	This information is formatted compactly. This information is presented concisely. This information is presented in a compact form. The representation of this information is compact and concise.	Lee et al. (2002)
14	Consistent representation	This information is consistently presented in the same format. This information is not presented consistently. This information is presented consistently. This information is represented in a consistent format.	Lee et al. (2002)
15	Ease of operation	This information is easy to manipulate to meet our needs. This information is easy to aggregate. This information is difficult to manipulate to meet our needs. This information is difficult to aggregate. This information is easy to combine with other information.	Lee et al. (2002)
16	Accurate & Objective	This information is objective, correct and accurate.	Lee et al. (2002)
17	Reliable & Trustwothy	This information is believable, credible, and reliable with a good reputation and comes from good sources. The Association of Computing Machinery's recommendation on open government (February 2009) stated, "published content should be digitally signed or include attestation of publication/creation date, authenticity, and integrity." digital signatures help the public validate the source of the data they find so that they can trust that the data has not been modified since it was published. Since provenance is for originally-published documents, it is not a reason to prevent the public from modifying government documents.	Lee et al. (2002)
18	Interpretability	It is easy to interpret what this information means. This information is difficult to interpret. It is difficult to interpret the coded information. This information is easily interpretable. The measurement units for this information are clear.	Lee et al. (2002)

(continued)

Table 8.5 (continued)

Dataset			
19	Timeliness	Data is made available as quickly as necessary to preserve the value of the data. This information is sufficiently current for our work. This information is not sufficiently timely. This information is not sufficiently current forour work. This information is sufficiently timely. This information is sufficiently up-to-date for our work.	Lee et al. (2002)
20	Understandability	This information is easy to understand. The meaning of this information is difficult to understand. This information is easy to comprehend. The meaning of this information is easy to understand.	Lee et al. (2002)
21	Delay in publication	Dataset: Indicates the ratio between the delay in the publication (number of days passed between the moment in which the information is available and the publication of the dataset) and the period of time referred by the dataset (week, month, year).	Vetrò, et al. (2016)
22	Delay after expiration	Dataset: Indicates the ratio between the delay in the publication of a dataset after the expiration of its previous version and the period of time referred by the dataset (week, month, year).	Vetrò, et al. (2016)
23	Comparability of today's data versus yesterday's data	Being able to rollback modification would allow historical analysis.	Lorenzo, Simone, Raimondo, and Federico (2015)

Table 8.6 Metrics – Metadata

Metadata			
1	Metadata availability	Documentation about the format and meaning of data goes a long way to making the data useful.	https://opengovdata.org/ (n.d.)
2	Title and description	Datasets should be provided together with their description and also how and for what purpose they were collected	Máchová and Lnénicka (2017)
3	Addressability & contactability	The extent to which the data publisher provide contact information. Addressability is another important dimension of open data since it emphasizes the extent to which contact information about the dataset's creator/maintainer is made available. Formally, the proposed metric defines the degree (%) to which datasets provide a value, an email address or HTTP URL to contact the data publisher [19].	Máchová and Lnénicka (2017)
4	Publisher	Datasets should be provided together with their publisher to verify authenticity of their source	Máchová and Lnénicka (2017)

(continued)

Table 8.6 (continued)

Metadata			
5	Release date and up to date	Datasets should be explicitly associated with a specific time or period tag. All information in the dataset should be up to date	Máchová and Lnénicka (2017)
6	Geographic coverage	Datasets should be determined if the coverage of data is on the national, regional or local level	Máchová and Lnénicka (2017)
7	Dataset URL	A URL must be provided in the metadata descriptions	Máchová and Lnénicka (2017)
8	Dataset (file) size	Datasets (file) size should be available	Máchová and Lnénicka (2017)
9	Number of views (visits)	Total number of online views should be available for a dataset	Máchová and Lnénicka (2017)
10	Number of downloads	Total number of downloads should be available for a dataset	Máchová and Lnénicka (2017)
11	Metadata completeness	Number of completed fields. The completeness metric deals with the number of completed fields in a metadata record. A meta-data record is considered complete, if the record contains all the information required to have an ideal representation of the described resource.	Reiche (2013)
12	Weighted completeness	Number of completed fields + weight. While the completeness metric is straightforward it comes with the drawback of treating every field with the same importance. The relevance of a certain metadata field depends strongly on the context. Not all fields might be relevant for the user when deciding whether the metadata record describes the resources he/she is looking for	Reiche (2013)
13	Metadata accuracy	The extent to which certain meta data values accurately describe the resources. Measures the semantic distance. The accuracy of a metadata record states whether the field values are correct with respect to the resources. In other words, how well does the metadata describe the actual resources?	Reiche (2013)
14	Richness of information	Measures the information content. The vocabulary terms and the description used in a metadata record should be meaningful to the user. For that the metadata need to contain enough information for describing uniquely the referred resource. From the user perspective, the metadata record is of high quality if he/she is confident enough about what the referenced resources contain	Reiche (2013)

(continued)

Table 8.6 (continued)

Metadata			
15	Metadata accessibility	Measures the readability. Accessibility measures the degree to which a metadata record is accessible in terms of cognitive accessibility, but also physical, respectively logical accessibility. The cognitive accessibility describe show easy a user can comprehend what the resource is about after reading the metadata record. In the matter of search ability this could decide, whether the user finds what he/she is looking for or not. Due to the domain-specific vocabulary of government it might be difficult to understand the description with ease. Thus, the readability might be an indicator for the general cognitive accessibility. To implement this metric several readability indexes could be used.	Reiche (2013)
16	Resource availability	Checks the availability of resources. With the availability not the metadata record itself is meant, but its resources. Metadata records define URLs which point to the actual resources. The availability metric assesses the number of reachable resources. A resource is available, if the resource can be retrieved. This could also mean, if the accessed page actually returns the described format. That would, however, rather be task of the accuracy metric. Different concerns are kept separated between different metrics	Reiche (2013)
17	Intrinsic precision	Number of spelling mistakes. The intrinsic precision is about the content of textual fields. Similar to the accessibility metric, this metric is about the reading fluency. The reading fluency is directly influenced by orthography of a text. Readers which are proficient in a language might halt for a moment on words written incorrectly. The number of spelling mistakes might not be a very important measure, as opposed to the availability of resources, nevertheless it influences the information quality.	Reiche (2013)
18	Track of creation	Dataset: Indicates the presence or absence of metadata associated with the process of creation of a dataset.	Vetrò et al. (2016)
19	Track of updates	Dataset: Indicates the existence or absence of metadata associated with the updates done to a dataset.	Vetrò et al. (2016)
20	Qr retrievability	The extent to which meta data and resources can be retrieved.	Umbrich, Neumaier, and Polleres (2015)
21	Qu usage	The extent to which available meta data keys are used to describe a dataset.	Umbrich et al. (2015)
22	Qc completeness	The extent to which the used meta data keys are non empty.	Umbrich et al. (2015)
23	Qo openness	The extent to which licenses and file formats conform to the open definition.	Umbrich et al. (2015)

Table 8.7 Metrics for linked data

Linked data			
1	COMP	Comparison between two literal values of a resource.	Kontokostas, Westphal and Auer (2014)
2	MATCH	The literal value of a resource matches/ does not match a certain regex pattern	Kontokostas, Westphal and Auer (2014)
3	LITRAN	The literal value of a specifically typed resource must (not) be within a given range	Kontokostas, Westphal and Auer (2014)
4	TYPEDEP	Type dependency: The type of a resource may imply the attribution of another type.	Kontokostas, Westphal and Auer (2014)
5	TYPRODEP	A resource of a specific type should have a certain property.	Kontokostas, Westphal and Auer (2014)
6	PVT	If a resource has a certain value V assigned via a property P1 that in some way classifies this resource, the existence of another property P2 can be assumed	Kontokostas, Westphal and Auer (2014)
7	TRIPLE	A resource can be considered erroneous if there are corresponding hints contained in the dataset	Kontokostas, Westphal and Auer (2014)
8	ONELANG	A literal value should contain at most one literal for a certain language	Kontokostas, Westphal and Auer (2014)
9	RDFSDOMAIN	The attribution of a resource's property (with a certain value) is only valid if the resource is of a certain type	Kontokostas, Westphal and Auer (2014)
10	RDFSRANGE	The attribution of a resource's property is only valid if the value is of a certain type	Kontokostas, Westphal and Auer (2014)
11	RDFSRANGED	The attribution of a resource's property is only valid if the literal value has a certain datatype	Kontokostas, Westphal and Auer (2014)
12	INVFUNC	Some values assigned to a resource are considered to be unique for this particular resource and must not occur in connection with other resources	Kontokostas, Westphal and Auer (2014)
13	OWLCARD	Cardinality restriction on a property	Kontokostas, Westphal and Auer (2014)
14	OWLDISJC	Disjoint class constraint	Kontokostas, Westphal and Auer (2014)
15	OWLDISJP	Disjoint property constraint	Kontokostas, Westphal and Auer (2014)
16	OWLASYMP	Asymmetric property constraint	Kontokostas, Westphal and Auer (2014)
17	OWLIRREFL	Irre exive property constraint	Kontokostas, Westphal and Auer (2014)

8.4.2 *System Quality*

System quality is divided into three dimensions; open data platforms capabilities dimension, the ease of use dimension and the performance dimension. When we are dealing with advanced Open Data platforms there could be one additional dimension referring to the data pro-cumers category of users; the data processing, enrichment and upload capabilities, which allows the users to further process the data upgrading them to more usable forms.

8.4.2.1 Open Data Platforms Capabilities

This category of evaluation metrics refers to the assessment of open data platforms capabilities. It could be used either from subjective (To what extend do you agree with the following statements? [7-point Likert scale]) or objective (Does the platform include the following functionality? [YES/NO]) models. It includes descriptive information about datasets and sources, functionalities provided by the Open Data portals in terms of dataset discovery, data provision capabilities, data visualization and multilingualism (Table 8.8).

8.4.2.2 Ease of Use

The ease of use metrics is forming a general dimension that could be used in the appraisal of every information system and service including open data platforms. These metrics are used mostly for subjective evaluation (Table 8.9).

8.4.2.3 Performance

The performance metrics is forming a general dimension that could be used in the appraisal of every information system and service including open data platforms. These metrics are used mostly for subjective evaluation but includes also metrics that could be used in objective evaluation (existence of API [YES/NO]) (Table 8.10).

8.4.2.4 Additional Dimension for Pro-Sumers

An additional dimension of evaluation metrics refer to the data procumers category of users as it is presented in Chap. 2. Data Processing and Upload Capabilities include functionalities provided by the open data portals in terms of enrichment, data cleansing, data linking and data format conversions. The pro–sumers concept was first introduced in (Charalabidis et al., 2014). It refers to subjects who concurrently provide and consume data and its quality. Subjects access the quality of data they

Table 8.8 Metrics for open data platforms capabilities

Search, provision and download			
1	Number of datasets	Portals should provide the number of datasets they include	Máchová and Lnénicka (2017)
2	Authority and responsibility	Portals should provide information about the authority, which hosts the portal and the governance model or institutional framework supporting data provision models	Máchová and Lnénicka (2017)
3	Number of applications (re-uses)	Portals should provide number of applications developed based on the open data re-used	Máchová and Lnénicka (2017)
4	Diversity of information	There are datasets from many different domains and/or countries	Charalabidis et al. (2014)
5	Thematic categories	PSI thematic categories: Economic and business information geographic information legal information meteorological and environmental information social information traffic and transport information tourist and leisure information agricultural, farming, forestry and fisheries information natural resources information	Alexopoulos et al. (2017)
6	RDF-compliance	It concerns the use of technologies that support RDF, including technical products of open data initiatives publishing structured data in a way that it can be interlinked, which as mentioned in the previous 'background' section is quite important for enabling more effective browsing and discovery of datasets, and for linking and combining OGD from multiple sources (e.g. see Villazón-Terrazas et al. (2011); Bauer and Kaltenböck (2012)); it is a binary indicator	Alexopoulos et al. (2015)
7	Download	The platform enabled me to download datasets easily and efficiently	Charalabidis et al. (2014)
8	Datasets discovery	It concerns the tools provided for discovering the datasets the user is interested in; its main possible values (not mutually exclusive) were: Simple document list, free text search, browsing through categories, browsing through filters, browsing through interactive map and SPARQL search.	Alexopoulos et al. (2017)
9	Visualizations	It concerns the datasets' visualization capabilities provided; one possible value is 'not existing', while other main possible values (not mutually exclusive) are visualizations in charts and visualizations in maps.	Alexopoulos et al. (2017)
10	Language	Portals should offer more language versions to gain more users (attention) and improve the overall quality of this portal	Máchová and Lnénicka (2017)

consume and are in position to mention weakness in them, and new needs they have. This concept eliminates the clear distinction between 'passive' content users/consumers and the 'active' content producers. In particular, next generation Open Data Infrastructures increasingly offer to data users capabilities for commenting and rating datasets, and also for processing them in order to improve them, adapt them to

Table 8.9 Metrics for ease of use

Ease of use			
1	Friendlyness	The platform provides a user friendly and easy to use environment	Charalabidis et al. (2014)
2	Easiness of use	It was easy to learn how to use the platform	Charalabidis et al. (2014)
3	Attractiveness	The web pages look attractive.	Charalabidis et al. (2014)
4	Design	It is easy to perform the tasks I want in a small number of steps.	Charalabidis et al. (2014)
5	Language adaptability	The platform allows me to work in my own language.	Charalabidis et al. (2014)
6	Personalisation	The platform supports user account creation in order to personalize views and information shown	Charalabidis et al. (2014)
7	Documentation	The platform provides high quality of documentation and online help.	Charalabidis et al. (2014)

their specialized needs, or link them to other datasets (public or private), and then uploading-publishing new versions of them, or even their own new datasets. In general, second generation of OGD infrastructures aim at fulfilling the needs of the emerging OGD 'pro-sumers' (Zuiderwick & Janssen, 2013) (Table 8.11).

8.4.3 Service Quality

Service quality consists of two dimensions; the license dimension and the feedback and collaboration dimension. When used for pro-cumers, it is expanded in the second one.

8.4.3.1 License

License dimension concerns license information related to the use of the published datasets. This is one of the most important characteristic of OGD sources, since it defines the allowed ways of OGD utilization and exploitation for generating various types of social and economic value, and reduces all relevant legal uncertainties (Table 8.12).

8.4.3.2 Feedback and Collaboration

Feedback and collaboration dimension concerns capabilities for users to communicate to the other users and the providers the level of quality of the datasets that they perceive. Also capabilities for users to get informed on the level of quality of the datasets perceived by other users through their ratings (e.g. five stars rating system).

Table 8.10 Metrics – performance

Performance			
1	Efficiency	The platform is always up and available without any interruptions.	Charalabidis et al. (2014)
2	Effectiveness	Services and pages are loaded quickly.	Charalabidis et al. (2014)
3	Bugs	I did not realize any bugs while using the platform.	Charalabidis et al. (2014)
4	API	Portals should provide API for stakeholders to develop applications using open data	Máchová and Lnénicka (2017)
5	Sources rating	According to the 5-stars Berner Lee's rating scheme for open data: *Make your stuff available on the web (whatever format) **Make it available as structured data (e.g. excel instead of image scan of a table) ***Using non-proprietary format (e.g. csv instead of excel) ****Use URLs to identify things, so that people can point at your stuff *****Link your data to other people's data to provide context	Alexopoulos et al. (2015)
6	Sources metadata rating	According to the 5-stars maturity scheme of metadata management. *Metadata ignorance **Scattered or closed metadata ***Open metadata for humans ****Open reusable metadata *****Linked open metadata	Alexopoulos et al. (2015)
7	Data management system	Portals should provide information about the data management system, which is used to power the portal	Máchová and Lnénicka (2017)
8	Social media	Portals should be connected to a social media platform to create a social distribution channel for open data. OGD users and providers can inform each other about what they did with and learned from a dataset	Máchová and Lnénicka (2017)
9	User account	Portals should support user account creation in order to personalize views and information shown	Máchová and Lnénicka (2017)

In addition it includes capabilities for users expressing their needs for additional datasets; getting informed about the needs of other users and getting informed about datasets extensions and revisions (Table 8.13).

8.5 Conclusions

The big investments made by governments of many countries for the development of OGD infrastructures, makes it necessary to evaluate them systematically, in order to understand better and assess the various types of value they generate, and identify

Table 8.11 Metrics for data pro-cumers

1	The platform provides good capabilities for data enrichment (i.e. adding new elements – fields)	Charalabidis et al. (2014)
2	The platform provides good capabilities for data cleansing (i.e. detecting and correcting ubiquities in a dataset)	Charalabidis et al. (2014)
3	The platform provides good capabilities for linking datasets.	Charalabidis et al. (2014)
4	The platform enabled me to upload datasets easily and efficiently.	Charalabidis et al. (2014)
5	The platform enabled me to prepare and add the metadata for the datasets I uploaded easily and efficiently.	Charalabidis et al. (2014)
6	The platform provides good capabilities for the automated creation of metadata.	Charalabidis et al. (2014)
7	The platform provides good capabilities for converting datasets' initial metadata in the metadata model of the platform easily and efficiently.	Charalabidis et al. (2014)
8	The platform provides strong API for uploading datasets (data and metadata	Charalabidis et al. (2014)

Table 8.12 Metrics for licencing

1	A presumption of openness	The presumption of openness rests on laws like the Freedom of Information Act, procedures including records management, and tools such as data catalogs.	https://opengovdata.org/ (n.d.)
2	Data license	It concerns license information related to the use of the published datasets	Alexopoulos et al. (2017)
3	Security	Provide information about restricted information. This information is protected against unauthorized access.	Lee et al. (2002)

the required improvements for increasing this value as it is presented in Chap. 7. In Chap. 3, we presented the major policies towards the achievement of this value. Policies should be evaluated measuring the impact of their developments. The expected impact of the Open Data policies and the development of data portals is to drive economic benefits and further transparency. These benefits have been largely outlined by a number of studies trying to develop evaluation models and metrics aiming at the assessment of those developments (impact and value assessment) as well as to drive next developments in the domain (maturity models, readiness assessment).

The studies have concentrated on the issues of open data quality assessment or the assessment of the portals offering them. The evaluation of an open data initiative or portal is a difficult task. Firstly, there are no objective and absolute (wide-accepted) metrics and targets (higher/lower values) for measurement. Secondly, there are too many perspectives for evaluation and each one of them provides different kind of insights.

As an evaluator, you first need to build the required evaluation model to fit your evaluation objectives. Then moving towards the finalization of the evaluation framework, a comprehensive evaluation procedure has to be developed for the use of the

Table 8.13 Metrics for feedback and collaboration

1	Quality rating	The platform provides good capabilities for giving feedback on the datasets I download, e.g. for rating datasets, for entering textual comments on them.	Charalabidis et al. (2014)
2	Feedback readability	The platform provides good capabilities for reading available feedback of other users of datasets I am interested in, e.g. ratings, comments.	Charalabidis et al. (2014)
3	Find users	The platform enables searching for and finding other users having similar interests with me in order to have information and knowledge ex-change and cooperation	Alexopoulos, Zuiderwijk, Charalabidis, Loukis, and Janssen (2016)
4	Groups of users	The platform enables forming groups with other users having similar interests with me in order to have information and knowledge exchange and cooperation	Alexopoulos et al. (2016)
5	Personalisation	The platform enables maintaining datasets/ working on datasets within one group	Alexopoulos et al. (2016)
6	Communication	The platform enables communicating with other users through messages in order to exchange information and knowledge	Alexopoulos et al. (2016)
7	Instant update	The platform enables getting immediately updated about the upload of new versions and enrichments of datasets maintained/worked on within the group, or new relevant items (e.g. publications, visualizations, etc.)	Alexopoulos et al. (2016)
8	Forum (feedback)	Portals should provide an opportunity to submit feedback on the data from the users to providers and forum to discuss and exchange ideas among the users	Máchová and Lnénicka (2017)
9	Request form	Portals should provide a form to request or suggest new type of format type of open data	Máchová and Lnénicka (2017)
10	Help	Portals should include high quality of documentation and help functionality to learn how to use the portal and improve the usability	Máchová and Lnénicka (2017)
11	Frequently Asked Questions (FAQ)	Portals should provide a FAQ section to help resolve any potential issues	Máchová and Lnénicka (2017)
12	Relevancy	This information is useful to our work. This information is relevant to our work. This information is appropriate for our work.	Lee et al. (2002)
Additional metrics for data pro-sumers in feedback and collaboration dimension			
13	Comments	The platform enables me to read interesting thoughts and ideas of the users on the datasets and the extensions I have uploaded by reading the comments they entered on them.	Alexopoulos et al. (2016)

(continued)

Table 8.13 (continued)

14	Rating	The platform enables me to get informed on the level of quality of the datasets and the extensions I have uploaded that is perceived by the users of them by reading their ratings	Alexopoulos et al. (2016)
15	Needs	The platform enables me to get informed about the needs of the users of the datasets and the extensions I have uploaded for additional ones	Alexopoulos et al. (2016)
16	Feedback	It concerns the existing tools allowing feedback from OGD users to the providers; its two main possible values were 'not existing' and 'existing'	Alexopoulos et al. (2015)

evaluation model (Alexopoulos et al., 2013). The procedure should include both quantitative and qualitative evaluation methods and tools to get deeper insights.

In this chapter we have presented quantitative models for objectively and subjectively evaluate an open data initiative. The metrics and models could be also used to develop tools for qualitative evaluation getting deeper insights by the end-users. Tools for qualitative evaluation such semi-structured questionnaires for discussion in a group of users, interviews and SWOT (Strengths-Weaknesses-Opportunities-Threats) analysis could be used for assessing various aspects of open data (impact, readiness, usability etc.).

A taxonomy of evaluation metrics has been developed in order to be used in alternative applications of the evaluation models based on the specific functionality of a platform or the quality of linked open data. Higher level models and tools have been presented towards the identification of the maturity and the evaluation of impact.

Chapter 9
Open Government Data: Areas and Directions for Research

"and still, so much to be done towards unveiling the true potential of open data."

9.1 Introduction

The concept of open data itself is strongly associated with innovative capacity and transformative power (Davies, Perini, & Alonso, 2013). It is increasingly recognized that proactively opening public data can create considerable benefits for several stakeholders, such as firms and individuals interested in the development of value added digital services or mobile applications, by combining various types of Open Government Data (OGD), and possibly other private data. On the other hand, OGD also empowers scientists, journalists and active citizens who want to understand various public issues and policies through advanced data processing and production of analytics (Janssen, 2011; Zuiderwijk, Helbig, Gil-García, & Janssen, 2014).

Due to its recognised potential to generate public value through driving innovation and economic growth, the OGD movement has been attracting a growing attention and interest of both researchers and practitioners from various disciplines, such as information systems, management sciences, political and social sciences and law. Research on open data has also been targeting the promoting of transparency and the substantiation of evidence-based decision making in policy formulation (Conradie & Choenni, 2012; Janssen, 2011; Stevens, 1984). At the same time, a few articles discussing unintended consequences and negative side effects of opening data have started to appear (Blakemore & Craglia, 2006; Zuiderwijk & Janssen, 2014a).

OGD, as a rather new organizational invention gradually diffusing in government is under a continuous renegotiation over its meanings and practices, and therefore a gradual formulation of its 'organizing vision', using the term proposed by Swanson and Ramiller (1997). According to Tammisto and Lindman (2012) the first level of renegotiation in the context of OGD took place initially in relevant

Y. Charalabidis et al., *The World of Open Data*, Public Administration and Information Technology 28, https://doi.org/10.1007/978-3-319-90850-2_9

policy discussions, public and professional press, and consultancy. The second level of renegotiation is taking place when organizations gradually understand how to benefit from open data and drive the development of social and economic value from it. This renegotiation and the evolution of this new domain can be greatly assisted by establishing a common code of understanding concerning the main areas and topics of research on OGD. However, despite the rapid growth of this multidisciplinary research domain, which has led to the emergence and continuous evolution of technologies and management approaches for open government data (OGD), a detailed analysis of the specific areas and topics of this research is still missing.

The development of a detailed taxonomy of current research areas and topics in the domain of OGD, presented in this Chapter, as part of the work done in (Charalabidis, Alexopoulos, & Loukis, 2016), will address the communication gap in this new domain, and facilitate better interaction among researchers and interested practitioners. It can also provide a solid base for driving future research in this domain, and thus contribute to reaching higher levels of maturity in the practices of opening and exploiting government data, as well as in the generation of greater social and economic value. The research taxonomy can assist in the development of a body of knowledge in this area, which will enable improving and optimizing the technology, the service design elements, the operations and overall performance of the units of government agencies responsible for opening data. Such a taxonomy is of critical importance for the development of a 'science base' (Charalabidis, Gonçalves, & Popplewell, 2011) in the OGD domain.

Research topics organisation is also extremely useful for Information and Communication Technology firms, assisting them in developing better OGD technological infrastructures, more innovative value added digital services or mobile applications based on OGD. This chapter contributes to filling the above-mentioned research gaps. In particular, it makes the following contributions:

(i) It develops a detailed taxonomy of research areas and corresponding research topics of the OGD domain has been developed, including four main research areas, which are further analysed into 35 research topics.
(ii) It comprises multi-sourced knowledge extraction process. The development of this taxonomy includes the extraction and combination of relevant knowledge originated from three different kinds of sources: important relevant government policy documents, research literature and experts from research and practice.
(iii) It ascertains these 35 research topics summarizing relevant research literature for each one of them. The main research objectives and directions have been highlighted and under-researched topics that require further research have been identified.
(iv) Our OGD research taxonomy extends and elaborates previous research taxonomies for the 'ICT-enabled Governance' and 'Policy Making 2.0' domains,

which have been developed in the FP7 European projects CROSSROAD and CROSSOVER.

(v) Finally, directions have been formulated for future multi-disciplinary research based on OGD aiming to address current societal challenges.

Part of the research presented in this chapter has been conducted within the FP7 ENGAGE project "An Infrastructure for Open, Linked Governmental Data Provision towards Research Communities and Citizens".

The chapter is structured as follows: Section 9.2 describes the methodology we followed for developing the taxonomy. In Sect. 9.3 the main findings of the literature review we have conducted for this purpose are presented and discussed. Then Sect. 9.4 presents the taxonomy, including descriptions of the identified main research areas, and the particular research sub-areas/topics for each of them. Finally, a discussion of findings is provided in Sect. 9.5, while Sect. 9.6 concludes the chapter.

9.2 Taxonomy Design Methodology

This study is focused on two main research questions, which constitute a first step towards the creation of a 'descriptive theory' of the OGD domain that will enable the development of a science base of it: (a) what are the main research areas and topics of the OGD domain, and (b) how they can be categorized? Gregor (2002) proposes five types of theories that need to be developed in the information systems domain; the first and more fundamental of them, which is necessary for the development of the other four more advanced ones, is the 'descriptive theories', which 'describe or classify specific dimensions or characteristics of individuals, groups, situations, or events'. There are two categories of descriptive theories: naming theories and classification theories (Stevens, 1984). A naming theory is a description of the main dimensions or characteristics of some phenomenon. A classification theory is more elaborate in that it also includes interrelations between such dimensions or characteristics of given phenomena.

This chapter contributes to the development of description theory for the OGD domain, both a naming and classification theory, which are of critical importance for the development of more advanced types of theories in this domain (e.g. concerning relationships between various dimensions or characteristics of them), and in general for the development of its scientific base. In particular, we developed an OGD research areas taxonomy, based on relevant government policy documents, previous relevant research literature and also experts' knowledge. For this purpose we followed the bottom-up approach to taxonomy development proposed by Ramos and Rasmus (2003) and Sujatha and Rao (2011), which includes the four stages shown in Fig. 9.1 (our research has focused on the first three of them).

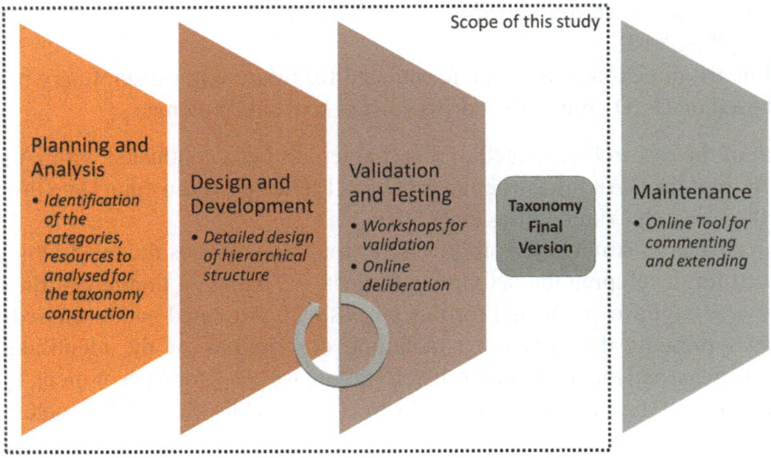

Fig. 9.1 The Open Data Research Taxonomy development approach

In particular, the methodology we followed for the development of the taxonomy was based on content analysis (Krippendorff, 2013) of different kinds of documents (government policy documents, previous relevant research literature and minutes of experts' workshops). It consisted of the following eight steps (shown also in Fig. 9.2):

1. Initially we identified and analysed important relevant government policy documents concerning OGD, which define the main terms, issues and perspectives, and also the main problems and challenges posed in this domain. The most important of them were:

   (a) European Commission Directives and Communications (European Commission, 2010a, 2011b, 2011d, 2012, 2013b, 2013d),
   (b) US Government documents (Executive Office of the President, 2009; Obama, 2012b),
   (c) UK Government documents (HM Government, 2012; O'Hara, 2011; UK Cabinet Office, 2011), and
   (d) Horizon 2020 Information and Communication Technologies Work Programme (European Commission, 2014).

The outcome of this step was a first set of ODG related terms, which were used for constructing the first version of the taxonomy, in step three.

2. Then we identified and analysed previous research papers that propose categorisations of research areas and perspectives of the OGD research domain. Additionally, we identified and analysed previous research literature concerning barriers to OGD publishing and exploitation, and also uptake of OGD and value generation from them. A brief review of this literature is presented in the following Sect. 9.3 The outcome of this step was another set of OGD related terms

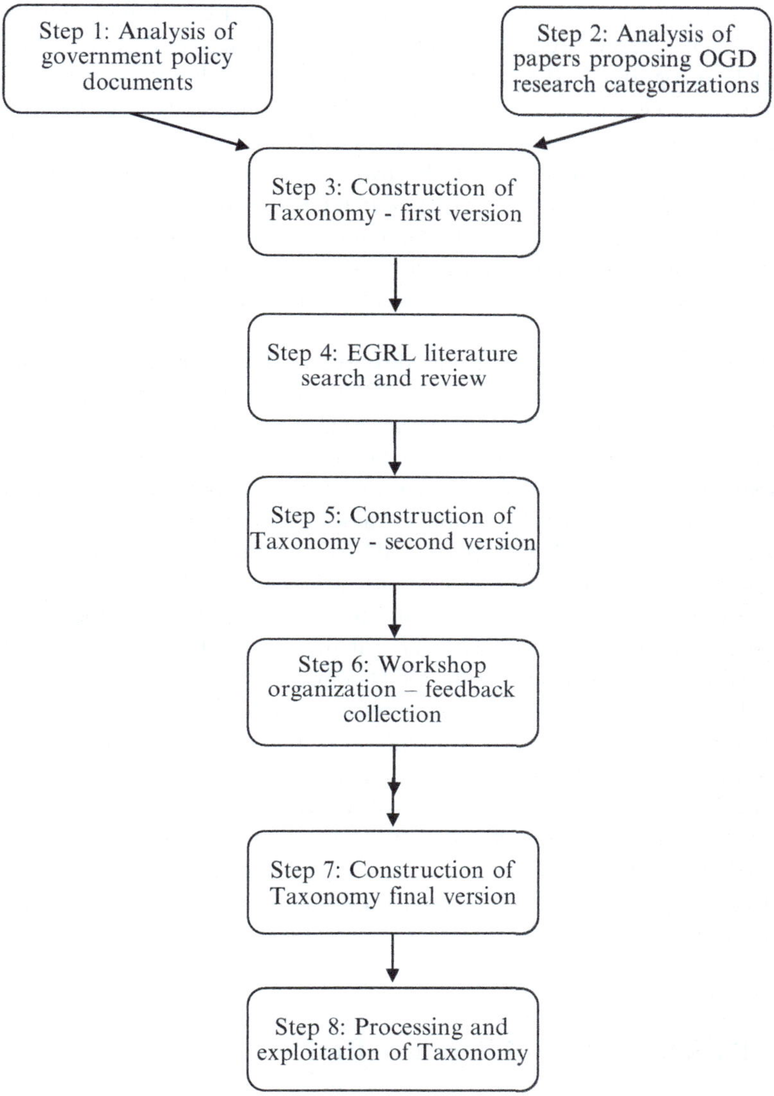

Fig. 9.2 Steps of the development methodology

(having some overlap with the ones of the set produced in the previous step), which were used as well for the construction of the first version of the taxonomy in step three.

3. After realising the above first two steps, the main research topics in the OGD domain were defined, and then were grouped in higher level research areas; this was a first version of the Open Data Research Taxonomy.

4. A thorough literature search was then conducted, based on the E-Government Reference Library (EGRL – faculty.washington.edu/jscholl/egrl/), which is a widely recognized and frequently updated electronic library of peer-reviewed papers in the electronic government/governance domain, using as keywords the terms of the above first version of the OGD Research Areas Taxonomy. In particular, the EGRL was searched by paper title and abstract for each of these terms, and the most relevant papers were retained and read in detail. This led to the identification of additional research topics in the OGD domain, which were used for the construction of a second version of the taxonomy.
5. The realisation of the fourth step resulted in the second version of the taxonomy.
6. A workshop was organised for the discussion, evaluation and validation of the above second version of the taxonomy, aiming at the assessment of its main research topics, and the possible proposition of new ones, and also at the assessment of their grouping, and the possible proposition of changes. Twenty OGD experts participated in this workshop in order to validate and further elaborate the second version of the taxonomy. These experts came from 11 different EU countries (NL, UK, DE, GR, BE, IT, AU, RO, ES, BG, LV), from different kinds of organizations (public administrations, universities and firms) and had different educational levels (Professors, PhD and MSc holders). All of the participants were selected based on their experience in the area of OGD and they are characterised as very experienced in the OGD domain, having been or currently being involved in OGD related projects (national or European).
7. Based on feedback collected from this workshop (which included the proposition of new research topics, such as topics 2.7 ('citizen-generated open data') and 2.8 ('sensor-generated open data') described in Sect. 9.4, and also of changes in their grouping in research areas), the final version of the taxonomy was produced, which is presented in Sect. 9.4.
8. Finally we proceeded to further processing and exploitation of it, and the results are presented in Sect. 9.5.

9.3 Background and Literature Review

During step 2 of the methodology as described in the previous section, we have identified four previous research papers that propose categorisations of OGD research in areas and themes (Davies et al., 2013; Harrison, Pardo, & Cook, 2012; Lindman, Rossi, & Tuunainen, 2014; Zuiderwijk et al., 2014), which were reviewed as they include elements that can be useful for the development of the Open Data Research Taxonomy.

The study of Davies et al. (2013, p.11) argues that *"over its short history as a field of action a number of distinct fronts of research into open data have developed, responding to different practice, policy and knowledge needs. These can be usefully classified into three broad groups: (1) open data readiness assessments, (2) open data implementation studies and (3) impact studies"*. Readiness studies aim to assess whether the conditions in public administrations are appropriate for the effective development of open data initiatives. Implementation studies aim to assess whether the conditions for open data itself actually exists in terms of open data availability, extent of publishing government agencies and importance of published datasets. Finally, impact studies aim to assess to what extent open data initiatives have led to change and public value.

The second study by Zuiderwijk et al. (2014, p.2) identifies seven different perspectives of OGD research, namely, (a) political, (b) social, (c) economical, (d) institutional, (e) operational, (f) legal and (g) technical and argues that *"combining perspectives may be more effective in dealing with the issues related to open data and stimulating innovation"*. Furthermore, it also identifies a number of OGD research directions, and categorises them under three major topics: (i) open data theory and development, (ii) open data policies, use, and innovation, and (iii) open data infrastructures and technologies.

Another study conducted by Lindman et al. (2014 p.4) focuses on the research challenges concerning Open Data Services, and categorises the relevant issues based on the work systems framework (Alter, 2010). It argues that *"there are two basic approaches for organizing the research issues according to the challenges that emerge when data is made available to the public, and further provided as services. These are: (1) an analysis of the life-cycle of the data and (2) an analysis of the levels of inquiry at which the open data phenomenon is studied"*. The proposed categories for the organisation of open data services research are: (1) Technologies, (2) Information, (3) Processes and Activities, (4) Products and Services, (5) Participants, (6) Customers and (7) Environment; each of them includes several research questions.

Finally, the study of Harrison et al. (2012, p.23) examines the Open Government 'ecosystem', concluding that OGD emerges as an essential dimension of the open government concept, arguing that *"the importance of developing the social and material infrastructures for creating, managing, and sharing data in the short term, along with the governance structures through which innovative architectures, infrastructures, and standards will be negotiated for the future"*. Then they define the main themes of the research required in order to realise this vision, along with the workflow of defining data of interest, prioritizing data collection, conducting data collection, publishing the data, and then using them and generating value.

Furthermore, there is another research stream dealing with the barriers to OGD publishing and exploitation (Barry & Bannister, 2014; Conradie & Choenni, 2012; Janssen, 2011; Janssen, Charalabidis, & Zuiderwijk, 2012; McDermott, 2010). We reviewed this research stream, as the main findings of it (e.g. identified barriers) might correspond to important research topics (e.g. concerning new ways of overcoming these barriers), so they can be useful for the development of the taxonomy. Finally, for the same reason we also reviewed another research stream dealing with the uptake and use of OGD, and their exploitation for innovation and value generation (Bason, 2010; Borins, 2001; Hartley, 2005; Kundra, 2012; Mohr, 1969; Windrum & Koch, 2008; Yang & Kankanhalli, 2013). The main conclusions of this stream of research indicate that the uptake and use of the OGD, and also the generation of innovation and value in general from them, are not straightforward, being complex, and requiring the collaboration of several actors.

From the above literature review we conclude that although there are some previous studies that propose categorisations of OGD research in areas and themes, they are at a too high level, and lack the detail required for directing future research. In order to provide the development of a 'science base' in this domain, we have to facilitate a better interaction among researchers and interested practitioners. Our research, as mentioned in the Introduction, contributes to filling this gap.

9.4 The Open Government Data Research Taxonomy

The Open Government Data Research Taxonomy consists of four major research areas (in its first level): OGD Management and Policies, OGD Infrastructures, OGD Interoperability and OGD Usage and Value (shown in Fig. 9.3), which include 35 research topics (in the second level). These 35 identified research topics were initially divided into two categories: the technological and non-technological ones; the latter correspond to the abovementioned OGD Usage and Value research area.

By examining the former we distinguished two clear sub-groups of research topics, concerning the interoperability and the management of the OGD respectively, which lead to the definition of the OGD Interoperability and the OGD Management and Policies areas; the remaining technological factors concerned the OGD infrastructures, so they were grouped in a separate research area. This grouping of the identified research topics into the above four research areas has been confirmed by the experts who participated in the workshop mentioned in the 'Methodology' Sect. 9.2. Changes were also proposed for some research topics and the research area they were associated with. The full taxonomy is available for reviewing and commenting online at mind42.com mind-mapping service[1].

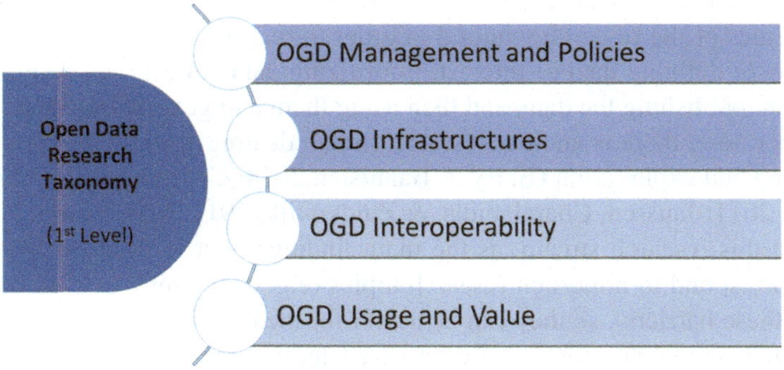

Fig. 9.3 Top-level open data research areas

[1] http://mind42.com/public/f2a7c2f6-63ec-475f-a848-7ed5abe6c5a4

9.4.1 *OGD Management and Policies*

The first top-level research area of the taxonomy has been named "Open Government Data Management and Policies". Data and information Management is an important research topic in the broader information systems domain, from which concepts, theories and frameworks can be borrowed and elaborated for further analysis and investigation of OGD management challenges.

Policy issues are closely related to the data management, in a broader definition, since policy decisions create the context of OGD management, so it affects data management procedures. Data management is a challenge both for OGD providers (public organizations) and for OGD users (e.g. scientists, analysts, journalists, active citizens). Therefore this research area includes several research topics corresponding to important OGD management challenges (such as methods for OGD anonymisation, cleansing, visualization, linking, publishing, mining, and also quality assessment). It is worth mentioning that within the workshop there were comments on whether we should put some of the research topics, such as OGD linking and mining in the category of infrastructures, since they are supported and provided by the developed infrastructures.

Finally, it was agreed that the OGD management capabilities, due to their importance for the use and the generation of value from OGD, should be viewed as a separate research area. In Fig. 9.4 we can see the research topics of the 'OGD Management and Policies' research area, while in Table 9.1 these OGD research topics are described in more detail, supported by some representative relevant literature from the EGRL.

OGD Management and Policies

- Policy and Legal Issues for OGD
- OGD Anonymisation Methods
- OGD Cleansing Methods
- OGD Quality Assessment Frameworks
- OGD Visualisation methods and tools
- OGD Linking
- OGD Publishing
- OGD Mining
- OGD Rating and Feedback

Fig. 9.4 Research topics for the OGD Management & Policies research area

Table 9.1 Description of the research topics of OGD Management & Policies research area

Research topic	Description
1.1 Policy & Legal Issues for OGD	This research topic concerns the investigation of different policies, strategies and principles for opening data, as well as specific measures and instruments in this direction (Blakemore & Craglia, 2006; European Commission, 2013b, 2013d; Zuiderwijk & Janssen, 2014b). Formulating an OGD policy is a complex multidisciplinary problem, and as such it is associated with many of the following research topics.
1.2 OGD Anonymisation Methods	The current practice in data publishing relies mainly on policies and guidelines as to what types of data can be published and on agreements concerning the use of published data. A major precondition for opening data of government agencies is not to disclose sensitive private data of citizens and firms. Therefore this research area focuses on methods for the anonymisation of opened data. Privacy-preserving data publishing (PPDP) provides methods and tools for publishing useful information while preserving data privacy (Fung, Wang, Chen, & Yu, 2010).
1.3 OGD Cleaning Methods	This research topic deals with data cleaning methods for OGD, which aim to correct errors in quantitative attributes of datasets, or even other types of attributes (Hellerstein, 2008). Data cleaning is a process used to determine inaccurate, incomplete or unreasonable data, and then improve their quality through correcting of detected errors and omissions. Generally data cleaning reduces errors and improves the data quality (Natarajan, Li, & Koronios, 2010).
1.4 OGD Quality Assessment Frameworks	This research topic deals with data quality, a major issue in information management in general, highly important for OGD in particular. Data quality problems occur anywhere in information systems, and they are solved by data cleaning (see previous research topic). After applying data cleaning, the quality of the data can be assessed in a number of ways, based on the internal consistency of the data and comparison of the corrected intensities with the corrected standard deviations (Chapman, 2005).
1.5 OGD Visualisation methods and tools	Visualization methods and tools is an important research topic, aiming to provide simple mechanisms for understanding and communicating large amounts of data. There is a need for exploratory mechanisms to navigate the data and metadata in these visualizations. It is therefore highly important to develop features and tools for facilitating the creation of visualizations by users on OGD (Graves & Hendler, 2013).
1.6 OGD Linking	The principles, frameworks, techniques and tools for OGD linking are the subjects of this research area (Bojārs, Breslin, Finn, & Decker, 2008; Kalampokis, Tambouris, & Tarabanis, 2013). The term linked data refers to data published on the web so that they are machine-readable, their meaning is explicitly defined, can be linked to (and from) other external datasets (Bizer, Heath, & Berners-Lee, 2009). The advancements on this research topic concentrate on how we can structure our data so that we can find, link and process them more easily. Knowledge management representation systems have been created and continue evolving in order to link different kinds of data.

(continued)

Table 9.1 (continued)

Research topic	Description
1.7 OGD Publishing	The OGD publishing research deals with and investigates all the issues of the publishing workflow and its involved actors (Bizer et al., 2009; Dawes & Helbig, 2010; Helbig, Cresswell, Burke, & Luna-Reyes, 2012). It also examines the interconnection between the OGD publishing processes and their context (main actors and their interests and goals), and also their effects on OGD use and outcomes, and on their dynamics.
1.8 OGD Mining	The OGD mining research aims to exploit and elaborate the algorithms and methods developed in the area of data mining, in order to extract useful patterns and knowledge from OGD. Data mining uses a broad family of computationally intensive methods which include decision trees, neural networks, rule induction, machine learning and graphic visualization (Bakirl et al., 2012; Mostafa & El-Masry, 2013).
1.9 OGD Rating and Feedback	This research focuses on policies and mechanisms for closing the feedback loop between OGD users and providers, through establishing communication channels between them (Zuiderwijk, 2015a). Another important objective of this research is to enable OGD providers to manage efficiently comments and requests from OGD users. Thus, tools for supporting the rating of OGD and their infrastructures, providing feedback to the corresponding public organizations are more than essential. The use of OGD users–providers collaboration techniques for the above purposes are also investigated in this research area, e.g. through web 2.0 oriented mechanisms (Alexopoulos, Zuiderwijk, Loukis, & Janssen, 2014; Charalabidis, Loukis, & Alexopoulos, 2014b).

9.4.2 OGD Infrastructures

The second research area of the Taxonomy has been named "Open Government Data Infrastructures". It includes research topics concerning various important technological aspects of the ICT infrastructures developed by government agencies in order to make OGD accessible to different groups of actors, such as their architectures, APIs provision and personalisation capabilities; another important research topic is OGD storage and long – term preservation, and also the use of cloud services in this domain.

Furthermore, though the main source of OGD is the information systems of government agencies, two more sources are gradually emerging, sensors and citizens, so researching them and their exploitation is an important research challenge. In Fig. 9.5 we can see the research topics of the 'OGD Infrastructure' research area, while in Table 9.2 these OGD research topics are described in more detail, supported also with representative literature from the EGRL.

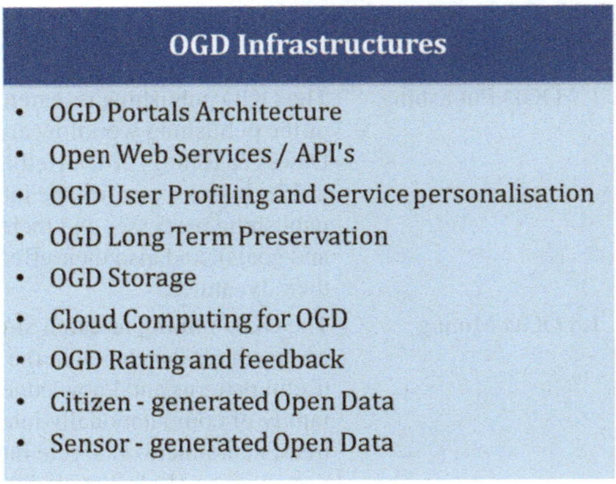

Fig. 9.5 Research topics for the OGD Infrastructures research area

Table 9.2 Description of the research topics of the OGD Infrastructures research area

Research topic	Description
2.1 OGD Portals Architecture	This research aims at defining the architectures of OGD portals, with respect to their scope and provided data and functionalities (Alexopoulos, 2016; Charalabidis et al., 2014b; Helbig et al., 2012). Various types and generations of architectures are proposed and discussed from various perspectives. Additionally, some research is conducted concerning the development of architectures of ICT infrastructures that allow for and support application development utilising OGD.
2.2 Open Web Services/APIs	This research aims at facilitating and providing well-designed standards for application programming interfaces (APIs) in OGD platforms, in order to ensure the exploitation and re-usability of published data. It is of high importance to use APIs for machine-to-machine operations for OGD. Unfortunately many of the OGD are not machine readable or the data are provided in a proprietary format (Braunschweig, Eberius, Thiele, & Lehner, 2012). Open web services in this domain should conform to a set of conventions that define how a client searches for and interacts with a service (Kleijnen & Raju, 2003; Paolucci, Kawamura, Payne, & Sycara, 2002).
2.3 OGD User Profiling and Service personalisation	This research focuses on user profiling, which can offer big opportunities to make OGD related services more personalised, to infer and predict citizens' behaviour, and to even influence their behaviour (Pieterson, Ebbers, & Dijk, 2005). Like the private sector, the public sector makes more and more use of user profiling in order to personalise the electronic services that are being offered to citizens (Mostafa & El-Masry, 2013).
2.4 OGD Long-term Preservation	This research topic can be found in every ICT related research domain, dealing with the ways and methods for the long-term preservation of data, which is particularly important for OGD (Agrawal & Srikant, 2000).
2.5 OGD Storage	This research topic concerns the optimization of OGD storage, combining knowledge from various domains, such as databases and algorithms.

(continued)

Table 9.2 (continued)

Research topic	Description
2.6 Cloud computing for OGD	The use of private and public cloud computing technologies and services (Lewis, 2013) for hosting and providing OGD is an important research challenge, taking into account the increasing adoption of cloud in the public sector (Joshi, 2012). The linked open data cloud creation supporting the vision of the web of data is also a research challenge classified under this research topic (Jain, Hitzler, Sheth, Verma, & Yeh, 2010; Jain, Hitzler, Yeh, Verma, & Sheth, 2010; Sorrentino, Bergamaschi, Fusari, & Beneventano, 2013).
2.7 Citizen-generated open data	This research aims to investigate the emerging and continuously growing volunteered user-generated content, which is often used to replace existing commercial or authoritative datasets, for example, Wikipedia[a] as an open encyclopaedia, or OpenStreetMap[b] as an open topographic dataset of the world (Richter & Winter, 2011) and Zooniverse[c] platform for people-powered research (many individual volunteers, relying on a version of the 'wisdom of crowds' to produce reliable and accurate data). Open data is generated by citizens, e.g. through e-participation platforms and social media, and their use for 'crowdsourcing' purposes, are an emerging research topic of this research area (Heipke, 2010).
2.8 Sensor-generated open data	This emerging research topic involves tools, methods and techniques for OGD generation through sensors, which will be made freely available to the public. Big data is becoming of critical importance for science and commercial applications development (e.g. Elgendy & Elragal, 2014b), so exploiting the knowledge developed in this domain and elaborating it for the OGD can be quite useful. This research topic also includes the development of methods of processing such data, calculation of analytics, and finally exploitation of them (for scientific and business purposes).

[a]http://en.wikipedia.org/wiki/Main_Page
[b]http://www.openstreetmap.org/
[c]https://www.zooniverse.org/

9.4.3 OGD Interoperability

Interoperability is a highly important feature of all types of information systems, and this gave rise to the development of a well-established research domain, which attracts considerable research interest, motivated by the increasing need of data exchange among organizations (both of the private and the public sector) (Jardim-Goncalves, Grilo, Agostinho, Lampathaki, & Charalabidis, 2013). Interoperability has many aspects, mainly technical, semantic and organisational. It becomes increasingly important in government, since *"The divergent interpretations of data, the lack of common metadata and the absence of universal reference data hinder governments from seamless data exchange, information systems integration and the delivery of cross-border public services"* (Shukair, Loutas, Peristeras, & Sklarss, 2013, p.10).

So our third research area deals with the interoperability issue in the specific domain of OGD. It includes research topics concerning OGD metadata, semantic

Fig. 9.6 Research topics for the OGD Interoperability research area

OGD Interoperability

- Metadata for OGD
- Multilinguality Issues
- Services Interoperability Standards
- Semantic Annotation
- OGD Ontologies
- Platform & Technical Interoperability
- Organisational Interoperability
- Controlled Vocabularies / Codelists Preservation

annotation, ontologies and controlled vocabularies and codelists, and also on OGD platforms technical interoperability, services interoperability standards and organizational interoperability. In Fig. 9.6 we can see the research topics of the 'OGD Interoperability' research area, which are described in more detail, and also supported with relevant literature from the EGRL, in Table 9.3.

9.4.4 *OGD Usage and Value*

The fourth research area of the research areas taxonomy is directed towards the measurement and deeper understanding of the use of OGD, as well as the impact and value generation from them. It includes research topics concerning on one hand OGD needs, readiness, use, skills management and reputation management, and on the other hand OGD related value and impact, innovation, entrepreneurship and contribution to accountability/transparency. In Fig. 9.7 we can see the research topics of this 'OGD Usage and Value' research area, while an elaboration of them and EGRL literature support are provided in Table 9.4.

9.5 Discussion

In this section the outcomes of the further processing and exploitation of the Research Areas Taxonomy are presented, conducted finally as part of the step eight of our research methodology (see Sect. 9.2): analysis of EGRL publications for each of the identified research topics (Sect. 9.5.1); exploitation of the Taxonomy for OGD Science Base Creation (Sect. 9.5.2); association of OGD Research Areas Taxonomy with the ICT-enabled Governance research taxonomy developed in the CROSSROAD and the CROSSOVER projects, and also use of the former in order to extend the latter (Sect. 9.5.3); and formulation of direction for multi-disciplinary research on important societal challenges using OGD (Sect. 9.5.4).

Table 9.3 Description of the research topics of the 'OGD Interoperability' research area

Research topic	Description
3.1 Metadata for OGD	This research topic includes various OGD metadata related research sub-topics: Data models, schemata, taxonomies, codelists and ontology-based extended metadata sets for OGD, and also other types e-government resources. The term semantic interoperability asset is widely used to refer to these types of resources (Charalabidis, Lampathaki, & Askounis, 2009; Robertson, Leadem, Dube, & Greenberg, 2001; Zuiderwijk, Jeffery, & Janssen, 2012b).
3.2 Multi-linguality	Multilinguality is a research topic that has been attracting a growing interest by supranational institutions, such as the European Union. It includes research associated with using, extending, combining and developing semantic assets towards the support of multi-linguality in the domain of OGD (Houssos, Jörg, & Matthews, 2012).
3.3 Service Interoperability Standards	This research topic concerns mainly the identification, composition and execution of various applications (designed and implemented independently) offered as services. This research investigates standards that can be used for seamless interconnection among OGD related services, in order to serve different OGD uses and user scopes (Jardim-Goncalves et al., 2013). It includes the development of information systems and registries consisting of workflow models and process descriptions in an integrated knowledge base (Sourouni, Lampathaki, Mouzakitis, Charalabidis, & Askounis, 2008).
3.4 Semantic Annotation	This research focuses on methods and tools for the semantic annotation of OGD generated by public organisations and sensors, as well as the semantic annotation of user-generated content (UGC) (Deng et al., 2013). Semantic annotation techniques capture not only the semantics, but also the pragmatics of the resources, such as who, when, where, how and why the resources are used (Dill et al., 2013; Kiryakov, Popov, Terziev, Manov, & Ognyanoff, 2004; Warner & Chun, 2009). The major objective of this research is the development of algorithms and tools for semantic integration (Bergamaschi, Castano, & Vincini 1999), and also for automated extraction of metadata (self-extracted metadata).
3.5 OGD Ontologies	This research topic includes investigation of the proper release of OGD and the use of ontologies behind these sources (Parundekar, Knoblock, & Ambite, 2010). Ontologies for the description and use of OGD, as well as the sense of ontology alignment are under investigation in this research (Osterwalder & Pigneur, 2010; Jain, Hitzler, Sheth et al., 2010; Jain, Hitzler, Yeh et al., 2010). The linked open data (LOD) paradigm is the major outcome of this research area.
3.6 Platform technical Interoperability	This research examines various technical issues involved in linking OGD systems and services, such as open interfaces, interconnection services, data integration, middleware, data presentation and exchange, accessibility and security services) (Jardim-Goncalves et al., 2013; Sarantis, Charalabidis, & Psarras, 2008).

(continued)

Table 9.3 (continued)

Research topic	Description
3.7 Organisational Interoperability	The main objective of this research is the investigation of the processes by which different organisations, such as different government agencies, collaborate in order to achieve mutually beneficial agreed e-government OGD service-related goals (Jardim-Goncalves et al., 2013; Sarantis et al., 2008), which concern the publishing and the management of OGD.
3.8 Controlled Vocabularies and Codelists Preservation	This research includes investigation regarding preservation, indexing, and retrieval of semantic assets, such as vocabularies and codelists (Kiryakov et al., 2004).

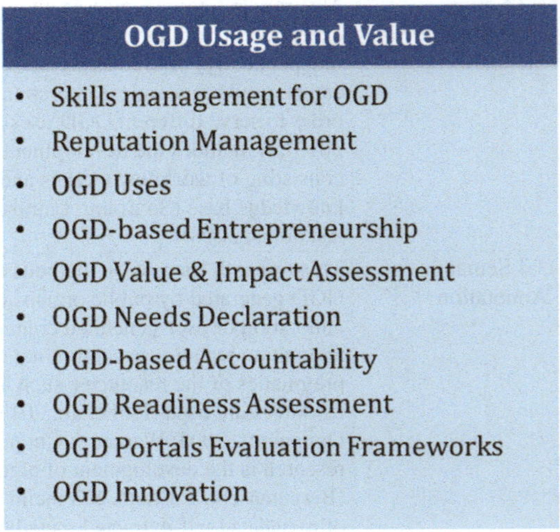

Fig. 9.7 Research topics for the OGD Usage and Value research area

Table 9.4 Description of the research topics of the 'OGD Usage and Value' research area

Research topics	Description
4.1 Skills Management for OGD	This research aims to identify and understand better the necessary skills required for OGD analysis and processing (by OGD users' side), and also for OGD publishing and management (by OGD providers' side). They are usually defined in terms of skills frameworks (also termed as competency frameworks or skills matrices); each of them consists of a list of skills, and a grading system, with a definition of what it means to be at particular level for a given skill.
4.2 Reputation Management	This research includes the investigation of the use of reputation systems in the OGD value chain. It examines various algorithms and methods for the reputation management of various OGD stakeholders (Bani & Paoli, 2013; Hansson, Verhagen, Karlstrom, & Larsson, 2013).

(continued)

Table 9.4 (continued)

Research topics	Description
4.3 OGD Use	It includes studies that describe and analyse examples, ways and paradigms of OGD use for various purposes, not only by citizens (e.g. scientists, journalists, active citizens, firms active in the development of value-added e-services and mobile applications), but also by the government (e.g. for policy making: Kalampokis, Hausenblas, and Tarabanis (2011), Kalampokis, Tambouris, and Tarabanis (2011b) combined social data and ODG for participatory decision-making in government).
4.4 OGD-based Entrepreneurship	This research topic concerns mainly business models for exploiting the potential value of OGD and initiating OGD value chains (Ferro & Osella, 2012, 2013).
4.5 OGD Value and Impact Assessment	The current OGD research on this topic focuses on analysing OGD initiatives that have led to the generation of some kind of public value (Charalabidis et al., 2014b; Davies et al., 2013; Jetzek, Avital, & Bjorn-Andersen, 2012, 2013), analysing the positive – and sometimes also the negative – aspects of OGD use and impacts.
4.6 OGD Needs Analysis	This research includes studies of OGD users' needs, with respect to both government datasets, and also functionalities of OGD infrastructures, aiming to lead to further developments of OGD strategies of public organizations, and also functionalities of ODG infrastructures/portals. For instance this research led to the identification of needs for collaboration workflows and feedback mechanisms (Alexopoulos et al., 2014), and also needs for better metadata and semantic annotation mechanisms (Zuiderwijk, 2015a).
4.7 OGD-based Accountability	This research investigates the use of OGD as part of anti-corruption programmes, in order to increase public sector accountability and credibility. Many government organizations publish a variety of datasets on the web, in order to promote transparency, accountability, and satisfy relevant legal obligations (Alon, 2011; Böhm et al., 2012b).
4.8 OGD Readiness Assessment	The main objective of this research is to develop frameworks and methods for assessing from various viewpoints (both 'internal' and 'external' ones) the degree of readiness of a national, regional or municipal government – or even individual agencies – to implement OGD initiatives (Davies et al., 2013; World Bank, 2013b).
4.9 OGD Portals Evaluation Frameworks	This research aims at the creation of roadmaps, guidelines and benchmarking frameworks for the evaluation of OGD portals and infrastructures from various viewpoints (Alexopoulos, 2016; Charalabidis et al., 2014b; Kalampokis et al., 2011).
4.10 OGD Innovation	The main objective of this OGD research is to identify and analyse innovations driven by OGD, both in the private sector (e.g. e-services innovations), and in the public sector (Zuiderwijk et al., 2014). According to this literature, OGD innovation concerns mainly three domains: (a) research, (b) business and (c) transparency (Jetzek et al., 2012, 2013). While the US literature and practice focuses mainly on (b), the EU tends to focus on (a), but both of them are equally interested in (c) OGD promotion towards transparency.

9.5.1 EGRL Publications for Research Topics

For all the OGD research areas identified and presented in the previous section (of the final version of the Taxonomy produced in step seven, see our methodology (Sect. 9.2)) we searched for relevant publications in the EGRL. In Fig. 9.8 we can see the number of publications found for each topic (the topics are sorted in descending order of publications' number); for the few publications that concern more than one of these topics we proceeded to their classification in the one judged as dominant (after discussion and consensus reaching among the authors).

We remark that there are significant differences among these research topics as to the number of relevant publications: for some of them we have found more publications, e.g. for research topics concerning OGD use, portals evaluation frameworks, publishing, policy and legal issues. For some others we found significantly less or even no publications, e.g. for research topics concerning sensor-generated OGD, OGD storage, long-term preservation, reputation management and skills management (for these five research topics there is no relevant literature in the EGRL. These topics were proposed in the workshop (step six of our OGD Research Areas Taxonomy development methodology by the experts who participated as major issues of OGD). Also, from Fig. 9.8 we can conclude that there are many under-researched topics with very small numbers of relevant publications. Therefore further research is required on these research topics with very small numbers or even no publications, since they constitute interesting emerging topics, which can be significant for the achievement of higher maturity in OGD practices and value generation from them.

9.5.2 Contribution to OGD Science Base Creation

As mentioned in Sect. 9.2, the research presented in this chapter contributes to the development of 'description theory' for the OGD domain, so it constitutes the first step towards the creation of a Science Base for it. According to Charalabidis, Gonçalves, and Popplewell (2010) the science base of a domain should include the main concepts, methods, tools and standards of the domain, and also supportive relevant experiments, surveys and case studies that have been conducted and produced a body of knowledge in the domain, and also various types of 'proofs of concept', aiming all to assist practitioners in this domain to solve particular problems and generate value.

Our OGD Research Areas Taxonomy contributes in the above-mentioned directions, as (i) it identifies the main concepts, methods and tools in OGD, and (ii) provides directions for future research in this domain, aiming to increase maturity of these methods and tools, so that finally OGD stakeholders (government, scientific communities, journalists, active citizens, and e–/m-services development firms) can be systematically assisted in their relevant activities, leading to higher value generation from OGD.

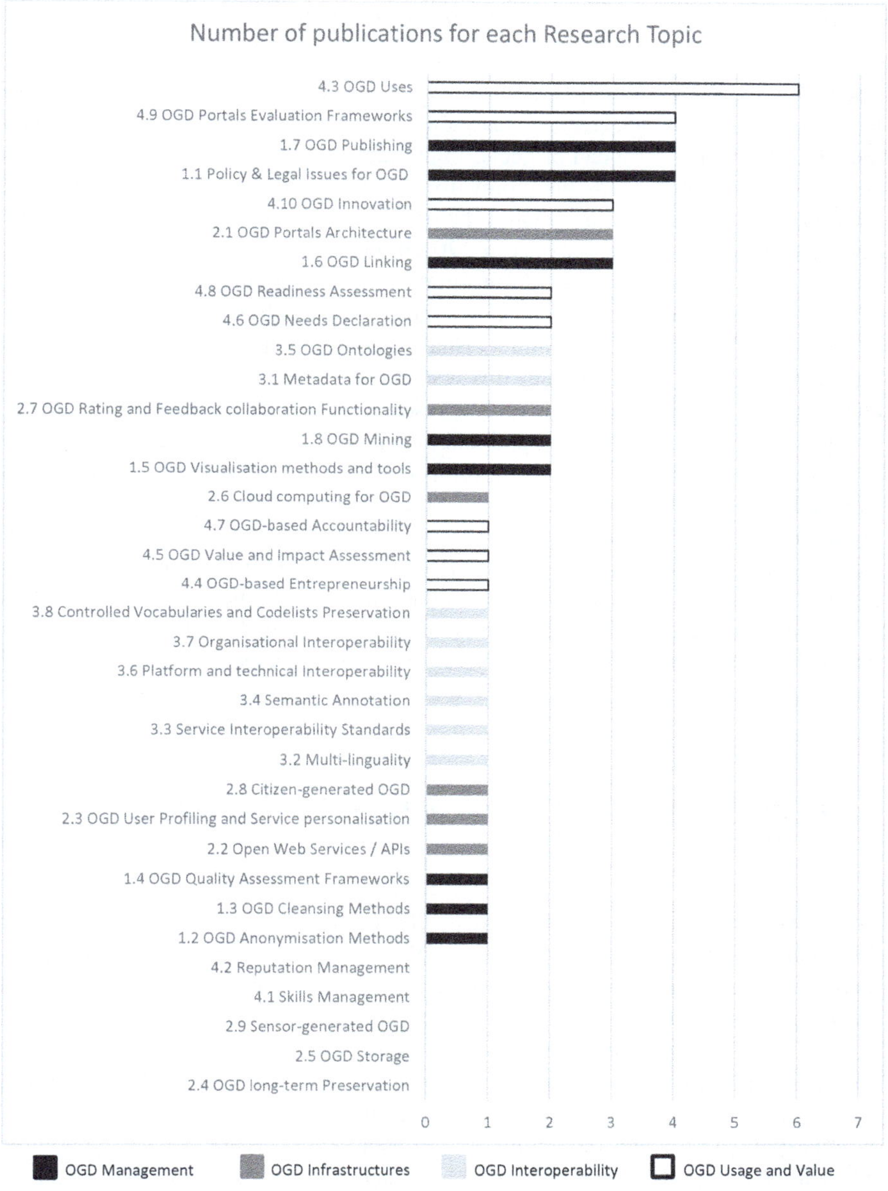

Fig. 9.8 Ranking of OGD research topics based on EGRL relevant literature

9.5.3 *Extension of ICT-Enabled Governance Taxonomy*

The OGD Research Areas Taxonomy is associated with and extends/elaborates the ICT-enabled Governance research taxonomy developed in the CROSSROAD[2] and the CROSSOVER[3] European projects. In particular, the CROSSROAD project has developed a research areas taxonomy for the ICT-enabled Governance domain, which consists of five main research themes, 17 research areas and more than 80 research sub-areas (Lampathaki et al., 2010). One of the research themes of this taxonomy is "Open Government Information & Intelligence for Transparency", which includes three research areas concerning "Open and Transparent Information Management", "Linked Data" and "Visual Analytics". The OGD Research Areas Taxonomy extends and elaborates this research theme, as the main research areas and topics of the former can replace the research areas and sub-areas of the latter, providing a higher level of detail and adding recently emerged research topics.

Also, the CROSSOVER project developed a taxonomy of research challenges in a related but narrower domain, concerning the next generation of public policy making in the Web 2.0 social media context (policy making 2.0) (CROSSOVER Project Deliverable 2.2.2, 2013), which categorises these research challenges under two research themes: (a) Data-powered Collaborative Governance and (b) Policy Modelling, in order to develop a roadmap on policy making 2.0. The OGD Research Areas Taxonomy extends and elaborates the "Linked Open Government Data" research challenge of the "Data-powered Collaborative Governance" theme.

9.5.4 *Multi-disciplinary Research on Societal Challenges Based on OGD*

In the workshops it was emphasized by the participating experts that the most important and socially beneficial research OGD research can be conducted by using them as a basis of multi-disciplinary research on important societal problems and challenges that modern societies face. These data can be used by multi-disciplinary scientific teams, e.g. including members from various 'neighbouring scientific domains', such as economic, political, social, management and behavioural sciences (and using theoretical foundations from these sciences) in order to perform various sophisticated analyses from various disciplinary perspectives and gain useful synthetic insights into serious problems and challenges of modern societies; these can be quite important for the design of effective solutions and public policies for addressing them. Some directions for such multi-disciplinary research were mentioned, and are summarized in Table 9.5.

---

[2] http://www.2020-horizon.com/CROSSROAD-CROSSROAD-A-Participative-Roadmap-for-ICT-Research-in-Electronic-Governance-and-Policy-Modelling(CROSSROAD)-s9412.html

[3] http://www.crossover-project.eu/ResearchRoadmap.aspx

Table 9.5 Directions of Multi-disciplinary Research on Societal Challenges Based on OGD

Societal challenge	ICT-enabled governance research topic	OGD research topic	Neighbouring scientific domain
Language divide and lack of cross-communities Communication	Language and cultural interoperability	Metadata for OGD Multilinguality Controlled vocabularies and CodelistsPreservation	Information intelligence Computer science (translation tools) Behavioural sciences
Anticipating unexpected crises	Social – economic simulation models Policy modelling Process optimization for OGD (accurate provision)	Semantic annotation Organisational interoperability Sensor-generated open data	Social and economic sciences
Enhanced collective cognitive intelligence (human/ICT-enabled) for better governance	Modelling and simulation Policy analysis Identity management	OGD mining Citizen-generated open data Visualization Information management	Economics Mathematics Sociology Computer science

9.6 Conclusions

As mentioned in the Introduction, the OGD research domain is still in its early stages, so it is important to develop a taxonomy of its main research areas and topics. The Open Government Data Research Taxonomy consists of four major research areas (in its first level): OGD Management and Policies, OGD Infrastructures, OGD Interoperability and OGD Usage and Value (shown in Fig. 9.3), which include 35 research topics (in the second level).

These 35 identified research topics have been validated through their association with relevant literature from the EGRL, as well as their importance to the experts of the workshop. The validation steps enabled a better understanding of them and their main research objectives and directions. Our OGD research taxonomy has been also connected with two previous research taxonomies for the 'ICT-enabled Governance' and 'Policy Making 2.0' domains respectively, which have been developed in the European projects CROSSROAD and CROSSOVER, providing extensions and elaborations of them for the OGD domain. Finally, directions have been formulated for future multi-disciplinary research based on OGD for addressing important challenges that modern societies face.

The findings of our study reveal the interesting thematic 'richness' of the OGD research domain, which currently includes a wide range of research topics, both technological and non-technological ones, concerning both the opening and publishing of government datasets, and also their usage (by various actors, such as e-service or mobile apps developers, scientists, analysts, journalists, active citizens, etc.), exploitation and value generation from them. This reflects the inherent complexity of opening of government data to the society and the economy, and then creating value from them, which the OGD research aims to address. In particular, we identified a multitude of technological research topics in the OGD research domain, with most of them concerning the exploitation of existing or emerging technologies, on one hand in the opened datasets (e.g. anonymisation, cleansing, mining, metadata, linking and semantically enriching technologies), and on the other hand in the OGD infrastructures (e.g. web services, storage, cloud computing, interoperability technologies), in order to enrich their usefulness. Furthermore, we identified a multitude of non-technological OGD research topics, which concern mainly OGD needs, use, impact, value and entrepreneurship.

Our study has revealed significant differences among the above identified OGD research topics as to the 'quantity' of the research conducted on them. For some of these topics there are limited or even no publications at all (e.g. for research topics sensor-generated OGD, OGD storage, long-term preservation, reputation management and skills management); so further research is required on these under-researched topics.

Our research taxonomy has interesting implications for research and practice. With respect to research it provides directions and structure for future research in the OGD domain, and also facilitates communication and interaction among researchers (through the 'common language' it introduces), and also with interested practitioners. Also, it contributes to the development of a 'description theory' of the OGD domain, which can be useful for the development of other more advanced types of theories (as mentioned in Sect. 9.2). Finally, it identifies important under-researched topics, on which further research is required. With respect to practice, the OGD Research Areas Taxonomy is useful to government agencies, as it proposes to them possible dimensions of their OGD strategies, practices and infrastructures, on which they should focus their attention, in order to improve the value generated from them. Also, this detailed taxonomy can contribute to the development of new knowledge in this domain, which will enable improving and optimizing the technology, and also the design, operations and performance of the units of government agencies responsible for opening data. Finally, the OGD Research Areas Taxonomy is useful to ICT firms developing OGD technological infrastructures, as it provides them directions for improving their products and services.

As the domain is evolving, it is necessary to organize more workshops in order to further validate the OGD Research Areas Taxonomy, and probably have proposals for additional research topics, with participants from all major stakeholder groups, such as such as e-service or mobile apps developers, scientists, analysts, journalists, active citizens and public servants. In this direction the proposed taxonomy is available on the Web and can be accessed through the following link: (http://mind42.com/public/f2a7c2f6-63ec-475f-a848-7ed5abe6c5a4), so that we can collect ratings, comments and ideas from the OGD community for further elaboration and update. Finally, it would be interesting to exploit other research libraries except EGRL and the multiple OGD research projects which are currently in progress (e.g. supported by European Commission or USA research programs) towards a better understanding of the implications in each research topic.

Appendix A: References

Acosta, M., Zaveri, A., Simperl, E., Kontokostas, D., Flöck, F., & Lehmann, J. (2018). Detecting Linked Data quality issues via crowdsourcing: A DBpedia study. *Semantic Web, 9*(3), 1–33.

Afuah, A. (2004). *Business models: A strategic management approach*. New York, NY: Irwin/McGraw-Hill.

Afuah, A., & Tucci, C. L. (2001). *Internet business models and strategies: Text and cases*. New York, NY: McGraw-Hill.

Agbabiaka, O., & Ojo, A. (2014). *Framework for assessing institutional readiness of government organisations to deliver open, collaborative and participatory services*. In Proceedings of the 8th International Conference on Theory and Practice of Electronic Governance (pp. 186-189). ACM.

Agrawal, R., & Srikant, R. (2000). *Privacy-preserving data mining*. In *Proceedings of the 2000 ACM SIGMOD international conference on Management of data (SIGMOD '00)*, ACM, (pp. 439–450). New York, NY. http://doi.acm.org/10.1145/342009.335438

Ajzen, I. (1991). The theory of planned behavior. *Organizational behavior and human decision processes, 50*(2), 179–211.

Alexopoulos, C. (2016). *Open government data infrastructures: research challenges, artefacts design and evaluation* (Doctoral dissertation, University of the Aegean. School of Science. Department of Information and Communication Systems Engineering). Karlovasi, Samos

Alexopoulos, C., Diamantopoulou, V., & Charalabidis, Y. (2017). *Tracking the evolution of OGD portals: A maturity model*.

Alexopoulos, C., Loukis, E., & Charalabidis, Y. (2014). A platform for closing the open data feedback loop based on Web2.0 functionality. *JeDEM, 6*(1), 62–68 Retrieved from http://www.jedem.org/article/view/327/270

Alexopoulos, C., Loukis, E., Charalabidis, Y., & Zuiderwijk, A. (2013). An evaluation framework for traditional and advanced open public data e-infrastructures. In W. Castelnovo, & E. Ferrari (Eds.). *Proceedings of the 13th European conference on Egovernment* (pp. 102–111). Como, Italy.

Alexopoulos, C., Loukis, E., Mouzakitis, S., Petychakis, M., & Charalabidis, Y. (2015). Analysing the characteristics of open government data sources in Greece. *Journal of the Knowledge Economy*, 1–33.

Alexopoulos, C., Zuiderwijk, A., Charalabidis, Y., Loukis, E., & Janssen, M. (2016). Designing a second generation of open data platforms: Integrating open data and social media. In *International Conference on Electronic Government* (pp. 230–241). Springer, Berlin, Heidelberg.

Y. Charalabidis et al., *The World of Open Data*, Public Administration and Information Technology 28, https://doi.org/10.1007/978-3-319-90850-2

Alexopoulos, C., Zuiderwijk, A., Loukis, E., & Janssen, M. (2014). *Designing a second generation of open data platforms: Integrating open data and social media, 2014*, Proceedings of EGOV 2014.

Algemene Rekenkamer. (2015). *Trendrapport open data 2015*. Retrieved from https://www.rekenkamer.nl/publicaties/rapporten/2015/03/31/trendrapport-open-data-2015

Algemene Rekenkamer. (2016). *Trendrapport open data 2016*. Retrieved from https://www.rekenkamer.nl/publicaties/rapporten/2016/03/24/trendrapport-open-data-2016

Ali- Eldin, A., Zuiderwijk, A., & Janssen, M. (2017). *Opening more data. A new privacy risk scoring model for open data*. Paper presented at the 7th International Symposium on Business Modeling and Software Design, Barcelona, Spain.

Allen, K. B. (1992). Access to government information. *Government Information Quarterly, 9*(1), 67–80.

Alon, P. (2011). When transparency and collaboration collide: The USA open data program. *Journal of the American Society for Information Science and Technology*, Wiley Subscription Services, Inc., A Wiley Company. https://doi.org/10.1002/asi.21622

Alter, S. (2010). Viewing systems as services: A fresh approach in the is field. *Communications of the Association for Information Systems, 26*(11), 2010.

Amit, R., & Zott, C. (2002). Value drivers of e-commerce business models. In M. A. Hitt, R. Amit, C. Lucier, & R. D. Nixon (Eds.), *Creating value: Winners in the new business environment* (pp. 15–47). Oxford, UK: Blackwell Publishers.

Anderson, C. (2009). *Free: The future of a radical price*. New York, NY: Hyperion Books.

Anderson, J. (1990). *Public policymaking: An introduction*. Boston, MA: Houghton Mifflin.

Andersen, K. V., & Henriksen, H. Z. (2006). E-government maturity models: Extension of the Layne and Lee model. *Government information quarterly, 23*(2), 236–248.

Andersen, K. N., Medaglia, R., & Henriksen, H. Z. (2012). Social media in public health care: Impact domain propositions. *Government Information Quarterly, 29*(4), 462–469.

Applegate, L. M. (2000). E-business models: Making sense of the internet business landscape. In G. Dickson & G. DeSanctis (Eds.), *Information technology and the future enterprise: New models for managers* (pp. 49–101). Englewood Cliffs, NJ: Prentice-Hall.

Arzberger, P., Schroeder, P., Beaulieu, A., Bowker, G., Casey, K., Laaksonen, L., … Wouters, P. (2004). Promoting access to public research data for scientific, economic, and social development. *Data Science Journal, 3*(29), 135–152.

Auer, S. (2011). *Creating knowledge out of interlinked data: Making the web a data washing machine*. Paper presented at the Proceedings of the International Conference on Web Intelligence, Mining and Semantics.

Auer, S., Bizer, C., Kobilarov, G., Lehmann, J., Cyganiak, R., & Ives, Z. (2007). DBpedia: A nucleus for a web of open data. In K. Aberer et al. (Eds.), *The semantic web. ISWC 2007, ASWC 2007. lecture notes in computer science* (Vol. 4825, pp. 722–735). Berlin, Heidelberg: Springer.

Auer, S., Bühmann, L., Dirschl, C., Erling, O., Hausenblas, M., Isele, R., … Williams, H. (2012). Managing the life-cycle of linked data with the LOD2 stack. In *International semantic Web conference* (pp. 1–16). Retrieved from http://svn.aksw.org/lod2/Paper/ISWC2012-InUse_LOD2-Stack/public.pdf

Auer, S., Lehmann, J., Ngomo, A.-C. N., & Zaveri, A. (2013). Introduction to linked data and its lifecycle on the web. In *Reasoning web. Semantic technologies for intelligent data access* (pp. 1–90). Heidelberg, Germany: Springer.

Bakirl, G., Birant, D., Mutlu, E., Kut, A., Denktaş, L., & Çetin, D. (2012). *Data mining solutions for local municipalities*. Paper presented at the 12th European conference on eGovernment (ECEG 2012), Barcelona, Spain.

Bani, M., & Paoli, S. D. (2013). *Ideas for a new civic reputation system for the rising of digital civics: Digital badges and their role in democratic process*. Paper presented at the 13th European conference on eGovernment (ECEG 2013), Como, Italy.

Barry, E., & Bannister, F. (2014). *Barriers to open data release: A view from the top*. In Proceedings 2013 EGPA annual conference, Edinburgh, Scotland, UK.

Bason, C. (2010). *"Leading public sector innovation", co-creating for a better society*. Bristol, UK: The Policy Press.

Batini, C., Cappiello, C., Francalanci, C., & Maurino, A. (2009). Methodologies for data quality assessment and improvement. *ACM Computing Surveys, 41*(3), 1–52. https://doi.org/10.1145/1541880.1541883

Bauer, F., & Kaltenbock, M. (2012). *Linked Open Data: The Essentials: A Quick Start Guide for Decision Makers*. edition mono/monochrom. Vienna, Austria, 23.

Behkamal, B., Kahani, M., Bagheri, E., & Jeremic, Z. (2014). A metrics-driven approach for quality assessment of linked open data. *Journal of theoretical and applied electronic commerce research, 9*(2), 64–79.

Bergamaschi, S., Castano, S., & Vincini, M. (1999). Semantic integration of semistructured and structured data sources. *ACM SIGMOD Record, 28*(1), 54–59.

Bernstein, M. S., Little, G., Miller, R. C., Hartmann, B., Ackerman, M. S., Karger, D. R., … Panovich, K. (2015). Soylent: A word processor with a crowd inside. *Communications of the ACM, 58*(8), 85–94.

Bertot, J. C., Jaeger, P. T., & Grimes, J. M. (2010). Using ICTs to create a culture of transparency: E-government and social media as openness and anti-corruption tools for societies. *Government Information Quarterly, 27*(3), 264–271.

Bizer, C., Heath, T., & Berners-Lee, T. (2009). Linked data – The story so far. *International Journal on Semantic Web, 5*(3), 1–22. https://doi.org/10.4018/jswis.2009081901

Blakemore, M., & Craglia, M. (2006). Access to public-sector information in Europe: Policy, rights and obligations. *The Information Society, 22*(1), 13–24.

Böhm, C., Freitag, M., Heise, A., Lehmann, C., Mascher, A., Naumann, F., … Schmidt, M. (2012a). *GovWILD: Integrating open government data for transparency*. Paper presented at the 21st International Conference Companion on World Wide Web, Lyon, France.

Böhm, C., Freitag, M., Heise, A., Lehmann, C., Mascher, A., Naumann, F., … Schmidt, M. (2012b). *GovWILD: integrating open government data for transparency*. In: Proceedings of the 21st international conference companion on World Wide Web (WWW '12 Companion). ACM, New York, NY, pp. 321–324. http://doi.acm.org/10.1145/2187980.2188039

Bojārs, U., Breslin, J. G. Finn, A., & Decker, S. (2008). Using the Semantic Web for linking and reusing data across Web 2.0 communities. *Web Semantics: Science, Services and Agents on the World Wide Web, 6*(1), 21–28. ISSN 1570-8268, https://doi.org/10.1016/j.websem.2007.11.010

Boley, H., & Chang, E. (2007). Digital ecosystems: Principles and semantics. In *Digital EcoSystems and Technologies conference, 2007. DEST'07. Inaugural IEEE-IES* (pp. 398–403). Retrieved from http://citeseerx.ist.psu.edu/viewdoc/download?doi=10.1.1.90.4199&rep=rep1&type=pdf

Borins, S. (2001). Encouraging innovation in the public sector. *Journal of Intellectual Capital, 2*(3), 310–319, 2001.

Borovina Josko, J. M., & Ferreira, J. E. (2017). Visualization properties for data quality visual assessment: An exploratory case study. *Information Visualization, 16*(2), 93–112.

Braunschweig, K., Eberius, J., Thiele, M., & Lehner, W. (2012). *The state of open data: Limits of current open data platforms*. Paper presented at the International World Wide Web Conference, Lyon, France. http://www2012.wwwconference.org/proceedings/nocompanion/wwwwebsci2012_braunschweig.pdf

Broad, E., Tennison, J., Starks, G., & Scott, A. (2015). *Who owns our data infrastructure?* Paper presented at the 3rd International Open Data Conference, Ottawa.

Brousseau, E., & Penard, T. (2006). The economics of digital business models: A framework for analyzing the economics of platforms. *Review of Network Economics, 6*(2), 81–110.

Bureau Woordvoering Kabinetsformatie. (2017). Vertrouwen in de toekomst. Regeerakkoord 2017 – 2021. VVD, CDA, D66 en ChristenUnie. Retrieved from https://www.kabinetsformatie2017.nl/documenten/publicaties/2017/10/10/regeerakkoord-vertrouwen-in-de-toekomst

Capgemini. (2015). *Creating value through open data: Study on the impact of re-use of public data resources*. European Commission. Brussels.

Carayannis, E. G., & Rakhmatullin, R. (2014). The quadruple/quintuple innovation helixes and smart specialisation strategies for sustainable and inclusive growth in Europe and beyond. *Journal of the Knowledge Economy, 5*(2), 212–239.

Carrara, W., Chan, W. S., Fischer, S., & van Steenbergen, E. (2015). *Creating value through open data*. European Union. https://doi.org/10.2759/328101.

Carrara, W., Fischer, S., & van Steenbergen, E. (2015). *Open data maturity in Europe 2015: Insights into the European state of play*. European Data Portal Open. CapGemini. Retrieved from https://www.capgemini.com/consulting/resources/open-data/

Carvalho P., Hitzelberger P., Otjacques B., Bouali F., Venturini G. (2015). Using Information Visualization to Support Open Data Integration. In M. Helfert, A. Holzinger, O. Belo, & C. Francalanci (Eds.) *Data Management Technologies and Applications. DATA 2014. Communications in Computer and Information Science*, (vol. 178, pp. 1–15). Springer, Cham.

Cavoukian, A. (2011). Privacy by design: Origins, meaning, and prospects for assuring privacy and trust in the information era. In G. O. M. Yee (Ed.), *Privacy protection measures and technologies in business organizations: Aspects and standards* (pp. 170–208). Aptus Research Solutions Inc. and Carleton University, Canada.

Chapman, A. D. (2005). *Principles and methods of data cleaning – primary species and species – occurrence data, version 1.0*. Report for the Global Biodiversity Information Facility, Copenhagen.

Charalabidis, Y., Alexopoulos, C., & Loukis, E. (2016). A taxonomy of open government data research areas and topics. *Journal of Organizational Computing and Electronic Commerce, 26*(1–2), 41–63 https://doi.org/10.1080/10919392.2015.1124720

Charalabidis, Y., Gonçalves, R. J., & Popplewell, K. (2010). Developing a science base for enterprise interoperability. In *Enterprise interoperability IV* (pp. 245–254). London, UK: Springer.

Charalabidis, Y., Gonçalves, R. J., & Popplewell, K. (2011). Towards a scientific foundation for interoperability. In Y. Charalabidis (Ed.), *Interoperability in digital public services and administration: Bridging E-government and E-business* (pp. 355–373). Hershey, NY: Information Science Reference.

Charalabidis, Y., Lampathaki, F., & Askounis, D. (2009). Metadata sets for e-government resources: The extended e-government metadata Schema (eGMS+). In M. A. Wimmer, H. J. Scholl, M. Janssen, & R. Traunmüller (Eds.), *Electronic government: 8th international conference (EGOV 2009)* (Vol. 5693, pp. 341–352). Berlin, Germany: Springer.

Charalabidis, Y., Loukis, E., & Alexopoulos, C. (2014). *Evaluating second generation open government data infrastructures using value models. In System Sciences (HICSS)*, 2014 47th Hawaii International Conference on (pp. 2114–2126). IEEE.

Chun, S. A., Shulman, S., Sandoval, R., & Hovy, E. (2010). Government 2.0: Making connections between citizens, data and government. *Information Polity, 15*(1/2), 1–9.

City of Chicago. (2012). *Open data executive order (no. 2012-2)*. Retrieved from https://www.cityofchicago.org/city/en/narr/foia/open_data_executiveorder.html

City of New York. (2016). *Open data policy and technical standards manual*. Retrieved from https://www1.nyc.gov/assets/doitt/downloads/pdf/nyc_open_data_tsm.pdf

Coglianese, C. (2009). The transparency president? The Obama administration and open government. *Governance, 22*(4), 529–544.

Cole, M., & Parston, G. (2006). *Unlocking public value: A new model for achieving high performance in public service organizations*. Hoboken, NJ: Wiley.

Committee on Earth Observation Satellites, Working Group on Information Systems and Services, U. S. G. S. (2011). *Data life cycle models and concepts*. Committee on Earth Observations Satellite. Retrieved from http://wgiss.ceos.org/dsig/whitepapers/Data%20Lifecycle%20Models%20and%20Concepts%20v8.docx

Conradie, P., & Choenni, S. (2012). *Exploring process barriers to release public sector information in local government*. Paper presented at the 6th international conference on theory and practice of electronic governance (ICEGOV), Albany, New York.

Cresswell, A. M., Burke, G. B., & Pardo, T. (2006). *Advancing return on investment, analysis for government IT: A public value framework*. Albany, NY: Center for Technology in Government, University at Albany.

CROSSOVER Project – Deliverable 2.2.2. (2013). *Towards policy – Making 2.0: The International roadmap on ICT for governance and policy modelling*. Retrieved from http://crossover-project.eu/Portals/0/0205F01_International%20Research%20Roadmap.pdf

da Silva Veith, A., dos Anjos, J. C. S., de Freitas, E. P., Lampoltshammer, T., & Geyer, C. F. (2016). Strategies for big data analytics through lambda architectures in volatile environments. *IFAC-PapersOnLine, 49*(30), 114–119. https://doi.org/10.1016/j.ifacol.2016.11.138

Daraio, C., Lenzerini, M., Leporelli, C., Naggar, P., Bonaccorsi, A., & Bartolucci, A. (2016). The advantages of an ontology-based data management approach: Openness, interoperability and data quality. *Scientometrics, 108*(1), 441–455.

Data.overheid.nl. (2017a). *Dataverzoek indienen*. Retrieved from https://data.overheid.nl/node/add/dataverzoek

Data.overheid.nl. (2017b). *Opvragen van informatie uit data.overheid.nl via de API*. Retrieved from https://data.overheid.nl/api

Data.overheid.nl. (2017c). *Over open data*. Retrieved from https://data.overheid.nl/over-open-data

Davies, T. (2013). *Open data barometer: 2013 global report*. Retrieved from http://www.opendataresearch.org/dl/odb2013/Open-Data-Barometer-2013-Global-Report.pdf

Davies, T., Perini, F., & Alonso, J. M. (2013). *Researching the emerging impacts of open data, ODDC conceptual framework*. Available at: http://www.opendataresearch.org/sites/default/files/posts/Researching%20the%20emerging%20impacts%20of%20open%20data.pdf

Davis, F. D. (1989). Perceived usefulness, perceived ease of use, and user acceptance of information technology. *MIS Quarterly, 13*(3), 319–339.

Davis, F. D., Bagozzi, R. P., & Warshaw, P. R. (1989). User acceptance of computer technology: a comparison of two theoretical models. *Management science, 35*(8), 982–1003.

Dawes, S., & Helbig, N. (2010). *Information strategies for open government: Challenges and prospects for deriving public value from government transparency*. Paper presented at the 9th international conference on e-government (EGOV), Lausanne, Switzerland.

Dawes, S. S., Vidiasova, L., & Parkhimovich, O. (2016). Planning and designing open government data programs: An ecosystem approach. *Government Information Quarterly, 33*(1), 15–27 https://doi.org/10.1016/j.giq.2016.01.003

De Vries, M., Kapff, L., Negreiro Achiaga, M., Wauters, P., Osimo, D., Foley, P., …, Whitehouse, D. (2011). *POPSIS – Pricing of public sector information study*. European Commission. http://ec.europa.eu/newsroom/dae/document.cfm?doc_id=1157

Debattista, J., Auer, S., & Lange, C. (2016). *Luzzu – A framework for linked data quality assessment*. Paper presented at the Semantic Computing (ICSC), 2016 IEEE Tenth International Conference on.

DeLone, D. H., & McLean, E. R. (1992). Information systems success: The quest for the dependent variable. *Information Systems Research, 3*(1), 60–95.

DeLone, D. H., & McLean, E. R. (2003). The DeLone and McLean model of information systems success: A ten-year update. *Journal of Management Information Systems, 19*(4), 9–30.

Demchenko, Y., Grosso, P., De Laat, C., & Membrey, P. (2013). Addressing big data issues in Scientific Data Infrastructure. In *Proceedings of the 2013 international conference on Collaboration Technologies and Systems, CTS 2013*. https://doi.org/10.1109/CTS.2013.6567203

Deng, D., Mai, G., Hsu, C., Chang, C., Chuang, T., & Shao, K. (2013). *Linking open data resources for semantic enhancement of user–Generated content*. Berlin/Heidelberg, Germany: Springer. 2013/01/01, https://doi.org/10.1007/978-3-642-37996-3_30

Dermeval, D., Vilela, J., Bittencourt, I. I., Castro, J., Isotani, S., Brito, P., & Silva, A. (2016). Applications of ontologies in requirements engineering: A systematic review of the literature. *Requirements Engineering, 21*(4), 405–437.

DG Connect. (2013). *A European strategy on the data value chain*, European Commission. http://ec.europa.eu/information_society/newsroom/cf/dae/document.cfm?doc_id=3488

Digital India. (n.d.). *Open government data (OGD) platform India – An overview*. Retrieved from http://meity.gov.in/writereaddata/files/OGD_Overview%20v_2.pdf

Dill, S., Eiron, N., Gibson, D., Gruhl, D., Guha, R., Jhingran, A., & Zien, J. Y. (2013). *SemTag and seeker: Bootstrapping the semantic web via automated semantic annotation*. In Proceedings of the 12th international conference on World Wide Web (pp. 178–186), ACM.

Directive, I. (2007). Directive 2007/2/EC of the European Parliament and of the Council of 14 March 2007 establishing an Infrastructure for Spatial Information in the European Community (INSPIRE). *Published in the official Journal on the 25th April*.

Dubosson-Torbay, M., Osterwalder, A., & Pigneur, Y. (2002). E-business model design, classification, and measurements. *Thunderbird International Business Review, 44*(1), 5–23.

Dutta B., Toulet A., Emonet V., Jonquet C. (2017) New Generation Metadata Vocabulary for Ontology Description and Publication. In E. Garoufallou, S. Virkus, R. Siatri, & D. Koutsomiha (Eds.) *Metadata and Semantic Research. MTSR 2017. Communications in Computer and Information Science*, (vol 755, pp. 173–185). Springer, Cham.

EC. (2011). *Communication from the Commission to the European Parliament, the Council, the European Economic and Social Committee and the Committee of the Regions Open data- An engine for innovation, growth and transparent governance. COM(2011) 882 final*. Brussels, Belgium: Commission of the European Communities.

Eisenmann, T., Parker, G., & Van Alstyne, M. W. (2006). Strategies for two-sided markets. *Harvard Business Review, 84*(10), 92–101.

Elgendy, N., & Elragal, A. (2014a). Big data analytics: A literature review paper. In P. Perner (Ed.), *Advances in data mining. Applications and theoretical aspects: 14th industrial conference, ICDM 2014, St. Petersburg, Russia, July 16–20, 2014. Proceedings* (pp. 214–227). Cham, Switzerland: Springer International Publishing.

Elgendy, N., & Elragal, A. (2014b). Big data analytics: A literature review paper. *Advances in Data Mining, Applications and Theoretical Aspects Lecture Notes in Computer Science, 8557*, 214–227.

Erl, T., Khattak, W., & Buhler, P. (2016). *Big data fundamentals: Concepts, drivers & techniques*. Boston, MA: Prentice Hall Press.

European Commission. (2003). *Directive 2003/98/EC of the European Parliament and of the council of 17 November 2003 on the re-use of public sector information*. Retrieved from http://ec.europa.eu/information_society/policy/psi/rules/eu/index_en.htm

European Commission. (2007). *Directive 2007/2/EC of the European Parliament and of the Council of 14 March 2007 establishing an Infrastructure for Spatial Information in the European Community (INSPIRE)*. Retrieved from http://eur-lex.europa.eu/LexUriServ/LexUriServ.do?uri=OJ:L:2007:108:0001:0014:EN:PDF

European Commission. (2010a). *Riding the wave: How Europe can gain form the rising tide of scientific data*. Brussels, Belgium.

European Commission. (2010b). *Communication from the Commission to the European Parliament and the Council Marine Knowledge 2020 marine data and observation for smart and sustainable growth*. Retrieved from http://eur-lex.europa.eu/legal-content/EN/ALL/?uri=CELEX:52010DC0461

European Commission. (2010c). *Directive 2010/40/EU of the European Parliament and of the Council of 7 July 2010 on the framework for the deployment of Intelligent Transport Systems in the field of road transport and for interfaces with other modes of transport Text with EEA relevance*. Retrieved from http://eur-lex.europa.eu/legal-content/EN/TXT/PDF/?uri=CELEX:32010L0040&from=EN

European Commission. (2011a). *Commission Recommendation of 27 October 2011 on the digitisation and online accessibility of cultural material and digital preservation*. Retrieved from http://eur-lex.europa.eu/LexUriServ/LexUriServ.do?uri=OJ:L:2011:283:0039:0045:EN:PDF

European Commission. (2011b). *Communication from the commission to the European parliament, the council, the European economic and social committee and the committee of the regions*, "Open data. An engine for innovation, growth and transparent governance", European Commission, Brussels, COM (2011) 882 final, 2011.

European Commission. (2011c). *Communication from the Commission to the European Parliament, the Council, the European Economic and Social Committee and the Committee of the Regions. Open data. An engine for innovation, growth and transparent governance*. Retrieved from Brussels: http://www.eu-spocs.eu/index.php?option=com_content&view=article&id=236:digital-agenda-turning-government-data-into-gold&catid=9:news&Itemid=56

European Commission. (2011d). *Digital agenda: Turning government data into gold*. European Commission, Brussels, P/11/1524, 2011.

European Commission. (2011e). *Digital agenda: Turning government data into gold*. Retrieved from http://europa.eu/rapid/press-release_IP-11-1524_en.htm?locale=en

European Commission. (2012, December). *Directive 2003/98/EC of the European parliament and of the council of 17 November 2003 on the re-use of public sector information*. European Commission. Available at: http://ec.europa.eu/information/society/policy/psi/rules/eu/ index en.htm.

European Commission. (2013a). *Commission welcomes Parliament adoption of new EU Open Data rules*. Retrieved from http://europa.eu/rapid/press-release_MEMO-13-555_en.htm

European Commission. (2013b). *EU implementation of the G8 open data charter*. Available: http://ec.europa.eu/information_society/newsroom/cf/dae/document.cfm?doc_id=3489

European Commission. (2013c). *Directive 2013/37/EU of the European Parliament and of the Council of 26 June 2013 amending Directive 2003/98/EC on the Re-use of Public Sector Information*. Retrieved from http://eur-lex.europa.eu/LexUriServ/LexUriServ.do?uri=OJ:L:2013:175:0001:0008:EN:PDF

European Commission. (2013d). *Digital agenda: Commission's open data strategy, questions & answers*. Available: http://europa.eu/rapid/pressReleasesAction.do?reference=MEMO/11/891&format=HTML&aged=1&language=EN&guiLanguage=en

European Commission. (2013e). *EU implementation of the G8 Open Data Charter*. Retrieved from http://ec.europa.eu/newsroom/dae/document.cfm?doc_id=3489

European Commission. (2014). Decision C (2014) 4995 of 22 July 2014. *HORIZON 2020 LEIT ICT Work Programme*. Available: http://ec.europa.eu/research/participants/data/ref/h2020/wp/2014_2015/main/h2020-wp1415-leit-ict_en.pdf

European Commission. (2016). *Report from the Commission to the Council and the European Parliament on the implementation of Directive 2007/2/EC of March 2007 establishing an Infrastructure for Spatial Information in the European Community (INSPIRE) pursuant to article 23*. Retrieved from http://eur-lex.europa.eu/legal-content/EN/TXT/?uri=CELEX%3A52016DC0478R%2801%29

European Commission. (2017). *European legislation on reuse of public sector information*. Retrieved from https://ec.europa.eu/digital-single-market/en/european-legislation-reuse-public-sector-information

European Data Portal. (2016a). *Netherlands – Overview*. Retrieved from https://www.europeandataportal.eu/sites/default/files/country-factsheet_netherlands.pdf

European Data Portal. (2016b). *Open data in Europe*. Retrieved from https://www.europeandataportal.eu/en/dashboard

European Data Portal. (2016c). *Open data maturity in Europe 2016. Insightsinto the European state of play*. Retrieved from https://www.europeandataportal.eu/sites/default/files/edp_landscaping_insight_report_n2_2016.pdf

European_Parliament_and_Council. (2003). Directive 2003/98/EC of 17 November 2003. On the re-use of public sector information. OJ L, 345, 90.

Evans, A. M., & Campos, A. (2013). Open government initiatives: Challenges of citizen participation. *Journal of Policy Analysis and Management, 32*(1), 172–185. https://doi.org/10.1002/pam.21651

Executive Office of the President. (2009). *Open government directive*. Available: http://www.whitehouse.gov/sites/default/files/omb/assets/memoranda_2010/m10-06.pdf

Faerman, S. R., McCaffrey, D. P., & Slyke, D. M. V. (2001). Understanding interorganizational cooperation: Public-private collaboration in regulating financial market innovation. *Organization Science, 12*(3), 372–388.

Farbey, B., Land, F., & Targett, D. (1999). Moving IS evaluation forward: Learning themes and research issues. *The Journal of Strategic Information Systems, 8*(2), 189–207.

Fassnacht, M., & Koese, I. (2006). Quality of electronic services. *Journal of Service Research, 9*(1), 19–37.

Ferro, E., & Osella, M. (2011). *Modelli di business nel riuso dell'informazione pubblica*, Report Osservatorio ICT della Regione Piemonte. http://www.osservatorioict.piemonte.it/it/images/

phocadownload/modelli%20di%20business%20nel%20riuso%20dellinformazione%20pubblica.pdf.

Ferro, E., & Osella, M. (2012). *Business models for PSI re-use: A multidimensional framework, using open data: Policy modeling, citizen empowerment*, Data Journalism Workshop, European Commission Headquarters, Brussels.

Ferro, E., & Osella, M. (2013). *Eight business model archetypes for PSI re-use, open data on the web workshop*, Google Campus, London.

Fishbein, M., & Ajzen, I. (1975). *Belief, attitude, intention and behavior: An introduction to theory and research.*

Fung, B. C. M., Wang, K, Chen, R., & Yu, P. S. (2010). Privacy-preserving data publishing: A survey of recent developments. *ACM Computing Surveys*. *42*(4), Article 14, 53 pages. http://doi.acm.org/10.1145/1749603.1749605

Gagliardi, D., Schina, L., Sarcinella, M. L., Mangialardi, G., Niglia, F., & Corallo, A. (2017). Information and communication technologies and public participation: Interactive maps and value added for citizens. *Government Information Quarterly, 34*(1), 153–166.

Gandomi, A., & Haider, M. (2015). Beyond the hype: Big data concepts, methods, and analytics. *International Journal of Information Management, 35*(2), 137–144 https://doi.org/10.1016/j.ijinfomgt.2014.10.007

Generalitat de Catalunya. (2017). *Partnership agreement between the Government of Catalonia and the Wikimedia Amical association*. Retrieved from http://dadesobertes.gencat.cat/web/.content/el_projecte_de_dades_obertes_gencat/acord_de_govern/convenis/2017013C_Acord-Viquipedia-SIGNAT.pdf

Gennari, J. H., Musen, M. A., Fergerson, R. W., Grosso, W. E., Crubézy, M., Eriksson, H., … Tu, S. W. (2003). The evolution of Protégé: An environment for knowledge-based systems development. *International Journal of Human-Computer Studies, 58*(1), 89–123.

Gerunov, A. (2016). Understanding open data policy: Evidence from Bulgaria. *International Journal of Public Administration, 40*(8), 649–657.

Governo Federal. (2010). *Manual Prático do Portal da Transparencia do Governo Federal*. Retrieved from http://www.portaltransparencia.gov.br/manual/manualCompleto.pdf

GovLab. (2014). *Welcome to the open data 500. Open data compass – What types of companies use which agencies' data?* Retrieved from http://www.opendata500.com/us/

Graves, A., & Hendler, J. (2013). *Visualization tools for open government data*. In proceedings of the 14th annual international conference on digital government research (dg.o '13), ACM, New York, NY, pp. 136–145. http://doi.acm.org/10.1145/2479724.2479746

Gregor, S. (2002). A theory of theories in information systems. In S. Gregor & D. Hart (Eds.), *Information systems foundations: Building the Theoretical Base* (pp. 1–20). Canberra, Australia: Australian National University.

Gruber, T. R. (1995). Toward principles for the design of ontologies used for knowledge sharing? *International Journal of Human-Computer Studies, 43*(5–6), 907–928.

Gruen, N., Houghton, J., & Tooth, R. (2014). *Open for business: How open data can help achieve the G20 growth target, (June)*, 14313.

Gunasekaran, A., Ngai, E. W. T., & McGaughey, R. E. (2006). Information technology and systems justification: A review for research and applications. *European Journal of Operational Research, 173*, 957–983.

Gyawali, B., Shimorina, A., Gardent, C., Cruz-Lara, S., & Mahfoudh, M. (2017). *Mapping natural language to description logic*. Paper presented at the European Semantic Web Conference.

Hammell, R., Bates, C., Lewis, H., Perricos, C., Brett, L., & Branch, D. (2012). Open data: Driving growth, ingenuity and innovation. *Deloitte White Pap.*

Hansson, K., Verhagen, H., Karlstrom, P., & Larsson, A. (2013). *Reputation and online communication: Visualizing reputational power to promote collaborative discussions*. Paper presented at the 46th Hawaii international conference on system sciences (HICSS-46), Wailea, HI.

Harrison, T. M., Guerrero, S., Burke, G. B., Cook, M., Cresswell, A., Helbig, N., … Pardo, T. (2012). Open government and e-government: Democratic challenges from a public value perspective. *Information Polity: The International Journal of Government & Democracy in the Information Age, 17*(2), 83–97. https://doi.org/10.3233/ip-2012-0269

Harrison, T. M., Pardo, T. A., & Cook, M. (2012). Creating open government ecosystems: A research and development Agenda. *Future Internet, 4*(4), 900–928 https://doi.org/10.3390/fi4040900

Hartley, J. (2005). Innovation in governance and public services: Past and present. *Public Money and Management, 25*(1), 27–34.

Hawkins, D. M. (2004). The problem of overfitting. *Journal of Chemical Information and Computer Sciences, 44*(1), 1–12.

Hazen, B. T., Overstreet, R. E., & Cegielski, C. G. (2012). Supply chain innovation diffusion: going beyond adoption. *The International Journal of Logistics Management, 23*(1), 119–134.

Heimstädt, M., Saunderson, F., & Heath, T. (2014). From toddler to teen: Growth of an open data ecosystem. *JeDEM – eJournal of eDemocracy and Open Government, 6*(2), 123–135 Retrieved from http://www.jedem.org/article/view/330

Heinrich, Bernd, Kaiser, Marcus und Klier, Mathias (2007) How to measure Data Quality? A Metric-based Approach. In *28th International Conference of Information Systems (ICIS), 2007*, Queen's University Montreal, canada.

Heipke, C. (2010). Crowdsourcing geospatial data. *ISPRS Journal of Photogrammetry and Remote Sensing, 65*(6), 550–557 ISSN 0924-2716, https://doi.org/10.1016/j.isprsjprs.2010.06.005

Helbig, N., Cresswell, A. M., Burke, G. B., & Luna-Reyes, L. (2012). *"The dynamics of opening government data", A white paper*. New York, NY: Center for Technology in Government, University at Albany, State University of New York.

Hellerstein, J. M. (2008). *Quantitative data cleaning for large databases*. United Nations economic Commission for Europe (UNECE).

Hitz, M., Kessel, T., & Pfisterer, D. (2017). *Towards sharable application ontologies for the automatic generation of UIs for dialog based linked data applications.* Paper presented at the MODELSWARD.

HM Government. (2012). *Open data white paper – Unleashing the potential*. Retrieved from http://data.gov.uk/sites/default/files/Open_data_White_Paper.pdf

Höchtl, J., & Lampoltshammer, T. J. (2016). ADEQUATe-Analytics and Data Enrichment to Improve the Quality of Open Data. In P. Parycek, & N. Edelmann (Eds.) *Proceedings of the International Conference for E-Democracy and Open Government CeDEM16*, (pp. 27–32). Edition Donau-Universität Krems, Krems.

Hofstede, G. (2001). *Culture's consequences. Comparing values, behaviors, institutions, and organizations across nations* (2nd ed.). Thousand Oaks, CA: Sage Publications.

Hofstede, G., Hofstede, G. J., & Minkov, M. (2010). *Cultures and organizations: Software of the mind* (3rd ed.). New York, NY: MCGraw-Hill.

Hofstede Insights. (2017). *Country comparison*. Retrieved from https://www.hofstede-insights.com/country-comparison/the-netherlands/

Hogan, A. (2013). Linked data and the semantic web standards. In A. Harth, K. Hose, & R. Schenkel (Eds.), *Linked data management* (pp. 3–48). Boca Raton, FL: CRC Press/Taylor & Francis.

Hogge, B. (2016). *Open corporates: Open data as a small part of the picture*. Omydiar Network. http://odimpact.org/files/case-study-open-corporates.pdf

Holsapple, C., Lee-Post, A., & Pakath, R. (2014). A unified foundation for business analytics. *Decision Support Systems, 64*, 130–141. https://doi.org/10.1016/j.dss.2014.05.013

Horrocks, I., Patel-Schneider, P. F., & Van Harmelen, F. (2003). From SHIQ and RDF to OWL: The making of a web ontology language. *Web Semantics: Science, Services and Agents on the World Wide Web, 1*(1), 7–26.

Houk, J. (2011). *Nike seeks fellow to start an open data revolution*. Retrieved from https://www.programmableweb.com/news/nike-seeks-fellow-to-start-open-data-revolution/2011/04/14

Houssos, N., Jörg, B., & Matthews, B. (2012). *A multi-level metadata approach for a Public Sector Information data infrastructure*. In Proceedings of the 11th International Conference on Current Research Information Systems (pp. 19–31).

Howe, J. (2006). The rise of crowdsourcing. *Wired Magazine, 14*(6), 1–4.

https://opengovdata.org/. (n.d.).
https://public.resource.org/8_principles.html. (2007). Retrieved from https://public.resource.org.
Huijboom, N., Van Den Broek, T., & Dutch Ministery of the Interior and Kingdom Relations. (2011). Open data: An international comparison of strategies. *European Journal of ePractice, 12*(April), 1–13 https://doi.org/1988-625X
IDC. (2017). *European data market study*. European Commission (Directorate-General for Communications Networks, Content and Technology). European Data Market. Ref.no.: SMART 2013/0063, Framingham, USA.
Irani, Z., & Love, P. (2008). Information systems evaluation – A crisis of understanding. In Z. Irani & P. Love (Eds.), *Evaluating information systems – Public and private sector*. Oxford, UK: Butterworth-Heinemann.
Jagadish, H. V., Gehrke, J., Labrinidis, A., Papakonstantinou, Y., Patel, J. M., Ramakrishnan, R., & Shahabi, C. (2014). Big data and its technical challenges. *Communications of the ACM, 57*(7), 86–94 https://doi.org/10.1145/2611567
Jain, P., Hitzler, P., Sheth, A. P., Verma, K., & Yeh, P. Z. (2010). Ontology alignment for linked open data. In *The semantic web–ISWC 2010* (pp. 402–417). Berlin/Heidelberg, Germany: Springer.
Jain, P., Hitzler, P., Yeh, P. Z., Verma, K., & Sheth, A. P. (2010). Linked data is merely more data. In D. Brickley, V. K. Chaudhri, H. Halpin, & D. McGuinness (Eds.), *Linked data meets artificial intelligence. Technical report SS-10-07* (pp. 82–86). Menlo Park, CA: AAAI Press ISBN 978-1-57735-461-1.
Janssen, K. (2011). *Legal interoperability – Barriers to the harmonization of licences*, presented at the ICRI – Share PSI workshop, Brussels.
Janssen, M., Charalabidis, Y., & Zuiderwijk, A. (2012). Benefits, adoption barriers and myths of open data and open government. *Information Systems Management, 29*(4), 258–268. https://doi.org/10.1080/10580530.2012.716740
Janssen, M., Estevez, E., & Janowski, T. (2014). Interoperability in big, open, and linked data – Organizational maturity, capabilities, and data portfolios. *IEEE Computer, 47*(10), 44–49.
Janssen, M., Matheus, R., Longo, J., & Weerakkody, V. (2017). Transparency-by-design as a foundation for open government. *Transforming Government: People, Process and Policy, 11*(1), 2–8.
Janssen, M., Matheus, R., & Zuiderwijk, A. (2015). *Big and open linked data (BOLD) to create smart cities and citizens: Insights from smart energy and mobility cases*. Paper presented at the EGOV2015: International Conference on Electronic Governmen, Thessaloniki, Greece.
Janssen, M., & Zuiderwijk, A. (2014). Infomediary business models for connecting open data providers and users. *Social Science Computer Review, 32*(5), 694–711 https://doi.org/10.1177/0894439314525902
Jardim-Goncalves, R., Grilo, A., Agostinho, C., Lampathaki, F., & Charalabidis, Y. (2013). Systematisation of interoperability body of knowledge: The foundation for enterprise interoperability as a science. *Enterprise Information Systems, 7*(1), 7–32.
Jeffery, K., Asserson, A., Houssos, N., & Jörg, B. (2013). *A 3-layer model for metadata*. Paper presented at the International Conference on Dublin Core and Metadata Applications, Lisbon, Portugal. http://dcevents.dublincore.org/IntConf/dc-2013/schedConf/presentations?searchField=&searchMatch=&search=&track=32
Jeffery, K., Houssos, N., Jörg, B., & Asserson, A. (2014). Research information management: The CERIF approach. *International Journal of Metadata, Semantics and Ontologies, 9*(1), 5–14.
Jetzek, T., Avital, M., & Bjorn-Andersen, N., (2012). *The value of open government data: A strategic analysis framework*, In: Proceedings of SIG eGovernment pre-ICIS Workshop, Orlando, USA.
Jetzek, T., Avital, M., & Bjørn-Andersen, N. (2013). The generative mechanisms of open government data. In *Proceedings of the 21st European Conference on Information Systems (ECIS 2013)*. Utrecht, The Netherlands.
Jonassen, D. H. (1991). Objectivism versus constructivism: Do we need a new philosophical paradigm? *Educational Technology Research and Development, 39*(3), 5–14.

Joshi, A. (2012). *Challenges for adoption of secured effective E-governance through virtualization and cloud computing*. Paper presented at the 9th international conference on E-governance (ICEG 2012), Cochin, Kerala, India.

Kalampokis, E., Hausenblas, M., & Tarabanis, K. (2011). Combining social and government open data for participatory decision-making. In E. Tambouris, A. Macintosh, & H. Bruijn (Eds.), *Electronic participation* (Vol. 6847, pp. 36–47). Berlin/Heidelberg, Germany: Springer.

Kalampokis, E., Tambouris, E., & Tarabanis, K. (2011a). Open government data: A stage model. *Lecture Notes in Computer Science, 6846*, 235–246.

Kalampokis, E., Tambouris, E., & Tarabanis, K. (2011b, 2011-01-01). *Open Government data: A stage model*. Berlin/Heidelberg, Germany: Springer. https://doi.org/10.1007/978-3-642-22878-0_20

Kalampokis, E., Tambouris, E., & Tarabanis, K. (2013). Linked open government data analytics. In M. A. Wimmer, M. Janssen, & H. J. Scholl (Eds.), *Electronic government* (pp. 99–110). Berlin/Heidelberg, Germany: Springer.

Kalampokis, E., Tambouris, E., & Tarabanis, K. (2017). ICT tools for creating, expanding and exploiting statistical linked open data. *Statistical Journal of the IAOS, 32*(2), 503–514.

Kalidien, S., Choenni, S., & Meijer, R. F. (2010). *Crime statistics online: Potentials and challenges*. Paper presented at the 11th Annual International Digital Government Research Conference on Public Administration Online: Challenges and Opportunities, Puebla, Mexico.

Karmanovskiy, N., Mouromtsev, D., Navrotskiy, M., Pavlov, D., & Radchenko, I. (2016). A case study of open science concept: Linked open data in university. In A. Chugonov, R. Bolgov, Y. Kabanov, G. Kampis, & M. Wimmer (Eds.), *Digital transformation and global society. DTGS 2016, Communications in computer and information science* (Vol. 674, pp. 400–403). Cham, Switzerland: Springer.

Kassen, M. (2013). A promising phenomenon of open data: A case study of the Chicago open data project. *Government Information Quarterly, 30*(4), 508–513. https://doi.org/10.1016/j.giq.2013.05.012

Kelly, G., Mulgan, G., & Muers, S. (2002). *Creating public value: An analytical framework for public service reform*. London, UK: UK Cabinet Office's Strategy Unit.

Kenya ICT Board. (2017). *Government of Kenya open data initiative*. Retrieved from https://www.google.nl/url?sa=t&rct=j&q=&esrc=s&source=web&cd=1&cad=rja&uact=8&ved=0ahUKEwj6tou34v7WAhVEaVAKHd4rCWQQFggnMAA&url=https%3A%2F%2Ffiles.ihub.co.ke%2Fihubresearch%2Fuploads%2F2012%2Faugust%2F1343900223__420.pdf&usg=AOvVaw0x4mAHYpATYmBh106e_569

Kifer, M. (2008). Rule interchange format: The framework. *RR, 8*, 1–11.

Kiryakov, A., Popov, B., Terziev, I., Manov, D., & Ognyanoff, D. (2004). Semantic annotation, indexing and retrieval. *Web Semantics: Science, Services and Agents on the World Wide Web, 2*(1), 49–79. ISSN 1570-8268, https://doi.org/10.1016/j.websem.2004.07.005

Kleijnen S., & Raju, S. (2003). An open web services architecture. *Queue, 1*(1), 38–46. http://dx.doi.acm.org/10.1145/637958.637961

Knap, T. (2017). Towards Odalic, a Semantic Table Interpretation Tool in the ADEQUATe Project. In A. L. Gentile, A. G. Nuzzolese, & Z. Zhang (Eds.), *Proceedings of the 5th International Workshop on Linked Data for Information Extraction co-located with the 16th International Semantic Web Conference (ISWC 2017)* (Vol. 1946, pp. 26–37).

Knap, T., Hanecák, P., Klímek, J., Mader, C., Necaský, M., Van Nuffelen, B., & Škoda, P. (2018). UnifiedViews: An ETL tool for RDF data management. *Semantic Web Journal*, pre-press, 1–16.

Konsti-Laakso, S. (2017). Stolen snow shovels and good ideas: The search for and generation of local knowledge in the social media community. *Government Information Quarterly, 34*(1), 134–139.

Kontokostas, D., Westphal, P., & Auer, S. (2014). Test-driven evaluation of linked data quality.

Kontokostas, D., Westphal, P., Auer, S., Hellmann, S., Lehmann, J., Cornelissen, R., & Zaveri, A. (2014a). *Test-driven evaluation of linked data quality*. Paper presented at the Proceedings of the 23rd International Conference on World Wide Web.

Kontokostas, D., Westphal, P., Auer, S., Hellmann, S., Lehmann, J., Cornelissen, R., & Zaveri, A. (2014b, April). Test-driven evaluation of linked data quality. In *Proceedings of the 23rd international conference on World Wide Web* (pp. 747–758). ACM. Seoul, Republic of Korea.

Koop, D., Santos, E., Mates, P., Vo, H. T., Bonnet, P., Bauer, B., … Silva, C. T. (2011). A provenance-based infrastructure to support the life cycle of executable papers. *Procedia Computer Science, 4*, 648–657 Retrieved from http://vgc.poly.edu/~juliana/pub/vistrails-executable-paper.pdf

Krippendorff, K. H. (2013). *Content analysis – an introduction to its methodology* (3rd ed.). Sage Publications. London, UK.

Krishnan, S., Teo, T. S., & Lim, V. K. (2013). Examining the relationships among e-government maturity, corruption, economic prosperity and environmental degradation: A cross-country analysis. *Information & Management, 50*(8), 638–649.

Krötzsch, M., Maier, F., Krisnadhi, A., & Hitzler, P. (2011). *A better uncle for OWL: Nominal schemas for integrating rules and ontologies*. Paper presented at the Proceedings of the 20th International Conference on World Wide Web.

Kucera, J. (2015). Open government data publication methodology. *Journal of Systems Integration*. https://doi.org/10.20470/jsi.v6i2.231

Kucera, J., & Chlapek, D. (2014). Benefits and risks of open government data. *Journal of Systems Integration*, 30–41 https://doi.org/10.20470/jsi.v5i1.185

Kulk, S., & van Loenen, B. (2012). Brave new open data world? *International Journal of Spatial Data Infrastructures Research, 7*, 196–206.

Kundra, V. (2012). *Digital fuel of the 21st century: Innovation through open data and the network effect*. Cambridge, MA: Joan Shorenstein Center on the Press, Politics and Public Policy.

Lampathaki, F., Charalabidis, Y., Passas, S., Osimo, D., Bicking, M., Wimmer, M., & Askounis, D. (2010). Defining a taxonomy for research areas on ICT for governance and policy modelling. In M. A. Wimmer, J.-L. Chappelet, M. Janssen, & H. J. Scholl (Eds.), *Electronic government, Lecture Notes in Computer Science* (Vol. 6228, pp. 61–72). Berlin, Germany: Springer.

Lampoltshammer, T. J., Guadamuz, A., Wass, C., & Heistracher, T. (2017). Openlaws.eu: Open justice in Europe through open access to legal information. In C. E. Jiménez-Gómez & M. Gascó-Hernández (Eds.), *Achieving open justice through citizen participation and transparency* (pp. 173–190). Hershey, PA: IGI Global.

Lampoltshammer, T. J., & Heistracher, T. (2014). Ontology evaluation with Protégé using OWLET. *Infocommunications Journal, 6*(2), 12–17.

Lampoltshammer, T. J., Sageder, C., & Heistracher, T. (2015). *The openlaws platform—An open architecture for big open legal data*. Paper presented at the Proceedings of the 18th International Legal Informatics Symposium IRIS.

Lampoltshammer, T. J., & Scholz, J. (2016). Citizen-driven geographic information science. In L. Ceccaron & J. Piera (Eds.), *Analyzing the role of citizen science in modern research* (pp. 231–243). Hershey, PA: IGI Global.

Lampoltshammer, T. J., & Scholz, J. (2017). Open Data as Social Capital in a Digital Society. In E. Kapferer, I. Gstach, A. Koch, & C. Sedmak (Eds.), *Rethinking social capital: Global contributions from theory and practice* (pp. 137–150). Newcastle upon Tyne, England: Cambridge Scholars Publishing.

Lampoltshammer, T. J., & Wiegand, S. (2015). Improving the computational performance of ontology-based classification using graph databases. *Remote Sensing, 7*(7), 9473–9491.

Lathrop, D., & Ruma, L. (2010). *Open government: Collaboration, transparency, and participation in practice*. Cambridge, MA: O'Reilly Media, Inc.

Layne, K., & Lee, J. (2001). Developing fully functional E-government: A four stage model. *Government information quarterly, 18*(2), 122–136.

Lee, D., Cyganiak, R., & Decker, S. (2014). *Open data Ireland: Best practice handbook*. Insight Centre for Data Analytics, NUI.

Lee, G., & Kwak, Y. H. (2012). An open government maturity model for social media-based public engagement. *Government Information Quarterly, 29*(4), 492–503.

Lee, Y. W., Strong, D. M., Kahn, B. K., & Wang, R. Y. (2002). AIMQ: a methodology for information quality assessment. *Information & management, 40*(2), 133–146.

Leimeister, J. M., Huber, M., Bretschneider, U., & Krcmar, H. (2009). Leveraging crowdsourcing: Activation-supporting components for IT-based ideas competition. *Journal of Management Information Systems, 26*(1), 197–224.

Lewis, G. A. (2013). *Role of standards in cloud-computing interoperability*, System Sciences (HICSS), 2013 46th Hawaii International Conference on, pp. 1652, 1661, 7–10 Jan 2013. https://doi.org/10.1109/HICSS.2013.470.

Linders, D. (2013). From e-government to we-government: Defining a typology for citizen coproduction in the age of social media. *Government Information Quarterly, 29*(4), 446–454.

Lindman, J., Rossi, M., & Tuunainen, V. K. (2014). *Open data services: Research agenda*, 2013/01/01, 2014 47th Hawaii International Conference on System Sciences. http://doi.ieeecomputersociety.org/10.1109/HICSS.2013.430

Lorenzo, C., Simone, B., Raimondo, L., & Federico, M. (2015). Collaborative open data versioning: A pragmatic approach using linked data. In *Conference for E-Democracy and Open Government* (p. 171–184).

Loukis, E., Pazalos, K., & Salagara, A. (2012). Transforming e-services evaluation data into business analytics using value models. *Electronic Commerce Research and Applications, 11*(2), 129–141.

Lu, J., & Zhang, G. (2003). Cost benefit factor analysis in e-services. *International Journal of Industry Service Management, 14*(5), 570–595.

Máchová, R., & Lnénicka, M. (2017). Evaluating the quality of open data portals on the national level. *Journal of theoretical and applied electronic commerce research, 12*(1), 21–41.

Malamud, C., O'Reilly, T., Elin, G., Sifry, M., Holovaty, A., O'Neil, D. X., … Newman, D. (2013). *Principles of open government data. Open Government Working Group*. Retrieved from http://www.opengovdata.org/home/8principles

Maslow, A. H. (1943). A theory of human motivation. *Psychological Review, 50*(4), 370–396.

McAfee, A., & Brynjolfsson, E. (2012). Big data: The management revolution. *Harvard Business Review, 90*, 60–66, 68, 128.

McBride, B. (2004). The resource description framework (RDF) and its vocabulary description language RDFS. In S. Staab & R. Studer (Eds.), *Handbook on ontologies* (pp. 51–65). Heidelberg, Germany: Springer.

McDermott, P. (2010). Building open government. *Government Information Quarterly, 27*(4), 401–413.

McLean, K. (2017). Smellmap: Amsterdam—Olfactory art and smell visualization. *Leonardo, 50*(1), 92–93.

Mihindukulasooriya N., García-Castro R., Priyatna F., Ruckhaus E., Saturno N. (2017). A Linked Data Profiling Service for Quality Assessment. In Blomqvist E., Hose K., Paulheim H., Ławrynowicz A., Ciravegna F., Hartig O. (Eds) *The Semantic Web: ESWC 2017 Satellite Events. ESWC 2014. Lecture Notes in Computer Science*, vol 10577, pp. 335-340. Springer, Cham

Miller, P., Styles, R., & Heath, T. (2008). Open data commons, a license for open data. *LDOW*, 369–374.

Minister of the Interior and Kingdom Relations. (2017). *Toepassing van de Wet openbaarheid van bestuur, Kamerstukken 32 802, nr. 37*. Retrieved from https://zoek.officielebekendmakingen.nl/kst-32802-37.html

Ministerie van Binnenlandse Zaken en Koninkrijksrelaties. (2013a). *Actieplan open overheid*. Retrieved from https://data.overheid.nl/sites/default/files/actieplan-open-overheid%20kopie.pdf

Ministerie van Binnenlandse Zaken en Koninkrijksrelaties. (2013b). *Visie open overheid*. Retrieved from https://data.overheid.nl/sites/default/files/visie-open-overheid%20kopie.pdf

Ministerie van Binnenlandse Zaken en Koninkrijksrelaties. (2016). *Nationale open data agenda 2016 (NODA) – Kamerstukken II 2014/15, 32 802, nr. 12*. Retrieved from https://www.rijksoverheid.nl/documenten/kamerstukken/2015/11/30/kamerbrief-over-nationale-open-data-agenda-2016-noda

Ministerie van Binnenlandse Zaken en Koninkrijksrelaties. (2017a). *Gemeentelijke high value Datalijst*. Retrieved from https://data.overheid.nl/gemeentelijke-high-value-datalijst

Ministerie van Binnenlandse Zaken en Koninkrijksrelaties. (2017b). *Jaarlijke data inventarisatie*. Retrieved from https://data.overheid.nl/data-inventarisatie

Ministerie van Binnenlandse Zaken en Koninkrijksrelaties. (2017c). *Open data beleid*. Retrieved from https://data.overheid.nl/open-data-beleid

Mohr, L. B. (1969). Determinants of innovation in organizations. *The American Political Science Review, 63*(1), 111–126.

Moller, K. (2013). Lifecycle models of data-centric systems and domains. *Semantic Web, 4*(1), 67–88 https://doi.org/10.3233/SW-2012-0060

Moon, M. J. (2002). The evolution of e-government among municipalities: Rhetoric or reality? *Public Administration Review, 62*(4), 424–433.

Morris, M., Schindehutte, M., & Allen, J. (2005). The entrepreneur's business model: Toward a unified perspective. *Journal of Business Research, 58*, 726–735.

Mostafa, M. M., & El-Masry, A. A. (2013). Citizens as consumers: Profiling e-government services' users in Egypt via data mining techniques. *International Journal of Information Management, 33*(4), 627–641. ISSN 0268-4012. https://doi.org/10.1016/j.ijinfomgt.2013.03.007.(http://www.sciencedirect.com/science/article/pii/S0268401213000510)

Natarajan, K., Li, J., & Koronios, A. (2010). *Data mining techniques for data cleaning*. London, UK: Springer https://doi.org/10.1007/978-0-85729-320-6_91

Nieuwenhuijs, S. (2014). *Het opvallendste nieuws volgens Sandor Nieuwenhuijs*. Retrieved from http://www.automatiseringgids.nl/nieuws/2014/05/het-opvallendste-nieuws-volgens-sandor-nieuwenhuijs

Nugroho, R. P., Zuiderwijk, A., Janssen, M., & de Jong, M. (2015). A comparison of national open data policies: Lessons learned. *Transforming Government: People, Process and Policy, 9*(3), 286–308.

Nurakmal, H., & Hamid, S. (2012). Post-adoption of open government data initiatives in public sectors.

O'Hara, K. (2011). *Transparent government, not transparent citizens: A report on privacy and transparency for the cabinet office, Gov. UK*, London, pp. 272–769.

Obama, B. (2009a). *Memorandum for the heads of executive departments and agencies: Transparency and open government*. Retrieved from https://www.whitehouse.gov/sites/whitehouse.gov/files/omb/memoranda/2009/m09-12.pdf

Obama, B. (2009b). *Open government directive*. Retrieved from http://www.whitehouse.gov/sites/default/files/omb/assets/memoranda_2010/m10-06.pdf

Obama, B. (2012a). *Digital government. Building a 21st century platform to better serve the American people*. Retrieved from https://obamawhitehouse.archives.gov/sites/default/files/omb/egov/digital-government/digital-government-strategy.pdf

Obama, B. (2012b). *Digital government. Building a 21st century platform to better serve the American people*. Available: http://www.whitehouse.gov/sites/default/files/omb/egov/digital-government/digital-government.html

ODB. (2016). *Open data barometer global report: Third edition*. http://opendatabarometer.org

OECD (2016), *Rebooting public service delivery: How can open government data help to drive innovation?* OECD Comparative Study.

Ojha S.R., Jovanovic, M., & Giunchiglia, F. (2015). Entity-Centric Visualization of Open Data. In Abascal J., Barbosa S., Fetter M., Gross T., Palanque P., Winckler M. (Eds) *Human-Computer Interaction – INTERACT 2015. INTERACT 2015. Lecture Notes in Computer Science*, (vol 9298, pp. 149–166). Springer, Cham

Ojo, A., & Adebayo, S. (2017). Blockchain as a next generation government information infrastructure: A review of initiatives in D5 countries. In A. Ojo & J. Millard (Eds.), *Government 3.0–Next generation government technology infrastructure and services* (pp. 283–298). Cham, Switzerland: Springer.

Ølnes, S. (2016). Beyond bitcoin enabling smart government using Blockchain technology. In H. J. Scholl, O. Glassey, M. Janssen, B. Klievink, I. Lindgren, P. Parycek, E. Tambouris, M. A. Wimmer, T. Janowski, & D. S. Soares (Eds.), *Proceedings of the 15th IFIP WG 8.5 international conference, EGOV 2016, Guimarães, Portugal, September 5-8, 2016* (pp. 253–264). Cham, Switzerland: Springer.

Ølnes, S., Ubacht, J., & Janssen, M. (2017). Blockchain in government: Benefits and implications of distributed ledger technology for information sharing. *Government Information Quarterly, 34*(3), 355–364. https://doi.org/10.1016/j.giq.2017.09.007

Olteanu, A., Ionita, A. D., & Solomon, A. S. (2017). *Curriculum and learning content management based on ontologies*. Paper presented at the The International Scientific Conference eLearning and Software for Education.

Open Data Charter. (2017). *History*. Retrieved from http://opendatacharter.net/history/

Open Data Monitor, P. (2015). *Data life cycle*. Retrieved from http://www.dataone.org/best-practices

Open Government Partnership. (2017). *About OGP*. Retrieved from https://www.opengovpartnership.org/about/about-ogp

Open Knowledge International. (2016). *Global open data index*. Retrieved from https://index.okfn.org/place/

Open Knowledge Network, P. (2017). *Advancing the state of open data through dialogue*. Open Knowledge Network. Retrieved from https://index.okfn.org/

Osterwalder, A. (2004). *The business model ontology: A proposition in a design science approach*, Dissertation 173, University of Lausanne, Switzerland.

Osterwalder, A., & Pigneur, Y. (2010). *Business model generation: a handbook for visionaries, game changers, and challengers*. John Wiley & Sons.

Otto, B., Jürjens, J., Schon, J., Auer, S., Menz, N., Wenzel, S., & Cirullies, J. (2016). *Industrial data space – digital Sovereignity over data*. Berlin, Germany: Fraunhofer-Gesellschaft zur Förderung der angewandten Forschung e.V Retrieved from https://www.fraunhofer.de/content/dam/zv/en/fields-of-research/industrial-data-space/whitepaper-industrial-data-space-eng.pdf.

Pan, J. Z. (2009). Resource description framework. In S. Staab & R. Studer (Eds.), *Handbook on ontologies* (pp. 71–90). Berlin Heidelberg, Germany: Springer.

Paolucci, M., Kawamura, T., Payne, T. R., & Sycara, K. (2002). *Semantic matching of web services capabilities*. Berlin/Heidelberg, Germany: Springer. https://doi.org/10.1007/3-540-48005-6_26. 2002-01-01.

Parasuraman, A., Zeithaml, V. A., & Berry, L. L. (1998). Alternative scales for measuring service quality: a comparative assessment based on psychometric and diagnostic criteria. In *Handbuch Dienstleistungsmanagement* (pp. 449–482). Wiesbaden: Gabler Verlag.

Parasuraman, A., Zeithaml, V. A., & Malhotra, A. (2005). E-S-QUAL: A multiple-item scale for assessing electronic service quality. *Journal of Service Research, 7*(3), 213–233.

Parundekar, R., Knoblock, C. A., & Ambite, J. L. (2010). Linking and building ontologies of linked data. In *The semantic web–ISWC 2010* (pp. 598–614). Berlin/Heidelberg, Germany: Springer.

Pazalos, K., Loukis, E., & Nikolopoulos, V. (2012). A structured methodology for assessing and improving e-services in digital cities. *Telematics and Informatics, 29*(1), 123–136.

Pedersen, T., Patwardhan, S., & Michelizzi, J. (2004). *WordNet:: Similarity: Measuring the relatedness of concepts*. Paper presented at the Demonstration papers at HLT-NAACL 2004.

Petticrew, M., & Roberts, H. (2008). *Systematic reviews in the social sciences: A practical guide*. Wiley.

Petychakis, M., Vasileiou, O., Georgis, C., Mouzakitis, S., & Psarras, J. (2014). A state-of-the-art analysis of the current public data landscape from a functional, semantic and technical perspective. *Journal of Theoretical and Applied Electronic Commerce Research, 9*(2), 34–47.

Pieterson, W., Ebbers, W., & Van Dijk, L. (2005). *The opportunities and barriers of user profiling in the public sector*. Berlin/Heidelberg, Germany: Springer. 2005-01-01, https://doi.org/10.1007/11545156_26

Pira International. (2010). *Commercial exploitation of Europe's public sector information*. European Commission Report, Surrey, England.

Piscini, E., Guastella, J., Rozman, A., & Nassim, T. (2016). Blockchain: Democratized trust – distributed ledgers and the future of value. In B. Briggs (Ed.), *Tech trends 2016 – Innovating in the digital era* (pp. 81–95). New York City, NY: Deloitte University Press.

Polleres, A. (2007). *From SPARQL to rules (and back)*. Paper presented at the Proceedings of the 16th International Conference on World Wide Web.

Pollitt, C., & Bouckaert, G. (2011). *Public management reform: A comparative analysis – New public management, governance, and the neo-weberian state*. Oxford, UK: Oxford University Press.

Pollock, R. (2011). *Building the (open) data ecosystem*. Open Knowledge Foundation Blog. Open Knowledge International Blog, .31. Retrieved from https://blog.okfn.org/2011/03/31/building-the-open-data-ecosystem/

Province Utrecht. (2017). *Utrecht open data*. Retrieved from http://www.utrechtopendata.org/

Quilitz, B., & Leser, U. (2008). *Querying distributed RDF data sources with SPARQL*. Paper presented at the European Semantic Web Conference.

Ramos, L., & Rasmus D. (2003). *Best practices in taxonomy development and management*. http://citeseerx.ist.psu.edu/viewdoc/summary?doi=10.1.1.201.4848

Reggy, L. (2011). Benchmarking open data availability across Europe: The case of EU structural funds. *European Journal of ePractice*. www.epracticejournal.eu N° 12, March/April. 2011.

Reiche, K. J. (2013). Assessment and visualization of metadata quality for open government data.

Richter, K. F., & Winter, S. (2011). *Citizens as database: Conscious ubiquity in data collection*. Berlin/Heidelberg, Germany: Springer. https://doi.org/10.1007/978-3-642-22922-0_27. 2011-01-01.

Robertson, W. D., Leadem, E. M., Dube, J., & Greenberg, J. (2001). Design and implementation of the National Institute of Environmental Health Sciences Dublin core metadata schema. In *International conference on Dublin core and metadata applications* (pp. 193–199).

Rowley, J. (2006). An analysis of the e-service literature: Towards a research agenda. *Internet Research, 16*(3), 339–359.

Ruijer, E., Grimmelikhuijsen, S., & Meijer, A. (2017). Open data for democracy: Developing a theoretical framework for open data use. *Government Information Quarterly, 34*(1), 45–52.

Saha, R., & Grover, S. (2011). Quantitative evaluation of website quality dimension for web 2.0. *International Journal of u- and e- Service, Science and Technology, 4*(4), 15–36.

Salguero, A., & Espinilla, M. (2018). A flexible text analyzer based on ontologies: an application for detecting discriminatory language. *Language Resources and Evaluation, 52*(1), 185–215.

Sarantis, D., Charalabidis, Y., & Psarras, J. (2008). Towards standardising interoperability levels for information systems of public administrations. *The Electronic Journal for E-commerce Tools & Applications (eJETA) Special Issue on Interoperability for Enterprises and Administrations Worldwide, 2*.

Savitz, A. W. (2006). *The Triple Bottom Line*. San Francisco: Jossey-Bass Wiley.

Schepers, J., & Wetzels, M. (2007). A meta-analysis of the technology acceptance model: Investigating subjective norm and moderation effects. *Information Management, 44*, 90–103.

Schroeder, M. (2008). Value theory. In E. N. Zalta (Ed.), *The Stanford encyclopaedia of philosophy*. Stanford, CA: Stanford University.

Second Chamber. (2015). *Kamerstukken II 2014/15, 34 123, nr. 13*. Retrieved from https://zoek.officielebekendmakingen.nl/kst-34123-3.html

Seddon, P. B. (1997). A respecification and extension of the DeLone and McLean model of IS success. *Information Systems Research, 8*(3), 240–253.

Seelos, C., & Mair, J. (2007). Profitable business models and market creation in the context of deep poverty: A strategic view. *Academy of Management Perspectives, 21*, 49–63.

Seničar, V., Jerman-Blažič, B., & Klobučar, T. (2003). Privacy-enhancing technologies—Approaches and development. *Computer Standards & Interfaces, 25*(2), 147–158. https://doi.org/10.1016/S0920-5489(03)00003-5

Shafer, S. M., Smith, H. J., & Linder, J. (2005). The power of business models. *Business Horizons, 48*, 199–207.

Shapiro, C., & Varian, H. R. (1999). *Information rules: A strategic guide to the network economy*. Boston, MA: Harvard Business School Press.

Share-PSI 2.0. (2016a). *Deliverable 7.2 stable version of the Share-PSI 2.0 best practices. Share-PSI 2.0 standards for open data and public sector information*. Retrieved from http://www.w3.org/2013/share-psi/bp/Share-PSI_D72

Share-PSI 2.0. (2016b). *Guides to implementation of the (revised) PSI directive*. Retrieved from https://www.w3.org/2013/share-psi/lg/

SHARE-PSI 2.0, P. (2016). *Deliverable 7.2 stable version of the share-PSI 2.0 best practices*. Online. Retrieved from https://www.w3.org/2013/share-psi/bp/Share-PSI_D72

Shuhaka, K., & Tauberer, J. (2012). *Business models for reuse of open legislative data*. Legalinformatics,

Shukair, G., Loutas, N., Peristeras, V., & Sklarss, S. (2013). Towards semantically interoperable metadata repositories: The Asset Description Metadata Schema. *Computers in Industry, 64*(1), 10–18. https://doi.org/10.1016/j.compind.2012.09.003

Smeulders, A. W., Worring, M., Santini, S., Gupta, A., & Jain, R. (2000). Content-based image retrieval at the end of the early years. *IEEE Transactions on Pattern Analysis and Machine Intelligence, 22*(12), 1349–1380.

Smith, A. (1776). *Of the origin and use of money: An inquiry into the nature and causes of the wealth of nations*. London, UK: W. Strahan.

Smith, B. (2003). Ontology. In L. Floridi (Ed.), *Blackwell guide to the philosophy of computing and information* (pp. 155–166). Oxford, UK: Blackwell.

Smithson, S., & Hirscheim, R. (1998). Analysing information systems evaluation: Another look at an old problem. *European Journal of Information Systems, 7*, 158–174.

Solar, M., Concha, G., & Meijueiro, L. (2012). A model to assess open government data in public agencies. In *International Conference on Electronic Government* (pp. 210–221). Springer, Berlin, Heidelberg.

Solar, M., Daniels, F., López, R., & Meijueiro, L. (2014). A model to guide the open government data implementation in public agencies. *Journal of UCS, 20*(11), 1564–1582.

Song, Y. (2017). Cross-Language Record Linkage Across Humanities Collections Using Metadata Similarities Among Languages. In: Kamps J., Tsakonas G., Manolopoulos Y., Iliadis L., Karydis I. (eds) *Research and Advanced Technology for Digital Libraries. TPDL 2017. Lecture Notes in Computer Science*, (vol 10450, pp. 640–643). Springer, Cham

Sorrentino, S., Bergamaschi, S., Fusari, E., & Beneventano, B. (2013). *Semantic annotation and publication of linked open data*. Berlin/Heidelberg, Germany: Springer https://doi.org/10.1007/978-3-642-39640-3_34

Sourouni, A. M., Lampathaki, F., Mouzakitis, S., Charalabidis, Y., & Askounis, D. (2008). Paving the way to eGovernment transformation: Interoperability registry infrastructure development. In *Electronic government* (pp. 340–351). Berlin/Heidelberg, Germany: Springer.

Soylu, A., Mödritscher, F., & De Causmaecker, P. (2012). Ubiquitous web navigation through harvesting embedded semantic data: A mobile scenario. *Integrated Computer-Aided Engineering, 19*(1), 93–109 https://doi.org/10.3233/ICA-2012-0393

Standaarden.overheid.nl. (2017). *Standaarden*. Retrieved from http://standaarden.overheid.nl/

State of New South Wales – Department of Finance, S. a. I (2016). *NSW government – Open data policy*. Retrieved from www.lsb.justice.nsw.gov.au/lsb/nswcopyright.html

Stevens, B. J. (1984). *Nursing theory. Analysis, application, evaluation* (2nd ed.). Boston, MA: Little, Brown.

Stewart, D. W., & Zhao, Q. (2000). Internet marketing, business models and public policy. *Journal of Public Policy and Marketing, 19*, 287–296.

Stewart, J., Jr., Hedge, D. M., & Lester, J. P. (2008). *Public policy: An evolutionary approach*. Australia: Thomson Wadsworth.

Straccia, U., & Bobillo, F. (2017). From fuzzy to annotated semantic web languages. In *Reasoning web: Logical foundation of knowledge graph construction and query answering* (pp. 203–240). Cham, Switzerland: Springer.

Stróżyna, M., Eiden, G., Abramowicz, W., Filipiak, D., Małyszko, J., & Węcel, K. (2017). A framework for the quality-based selection and retrieval of open data-a use case from the maritime domain. *Electronic Markets 28*(2), 219–233.

Sugimoto, S., Li, C., Nagamori, M., & Greenberg, J. (2017). Permanence and temporal interoperability of metadata in the linked open data environment. *International Conference on Dublin Core and Metadata Applications* DC-2017, (pp. 45–54), Washington, D.C.

Sujatha, R., & Rao, B. R. K. (2011). Taxonomy construction techniques–issues and challenges. *Indian Journal of Computer Science and Engineering, 3*, 5.

Sumak, B., Polancic, G., & Hericko, M. (2009). *Towards an e-service knowledge system for improving the quality and adoption of e-services*. In Proceedings of the 22nd Bled 'eEnablement: Facilitating an Open, Effective and Representative Society', June 14–17, 2009, Bled, Slovenia.

Sunlight Foundation. (2014). *Open data policy guidelines*. Retrieved from https://sunlightfoundation.com/opendataguidelines/

Susha, I., Janssen, M., & Verhulst, S. (2017). Data collaboratives as "bazaars"?: A review of coordination problems and mechanisms to match demand for data with supply. *Transforming Government: People, Process and Policy, 11*(1), 157–172 https://doi.org/10.1108/TG-01-2017-0007

Susha, I., Zuiderwijk, A., Charalabidis, Y., Parycek, P., & Janssen, M. (2015). Critical factors for open data publication and use: A comparison of city-level, regional, and transnational cases. *eJournal of eDemocracy and Open Government, 7*(2), 94–115.

Susha, I., Zuiderwijk, A., Janssen, M., & Gronlund, A. (2014). Benchmarks for evaluating the progress of open data adoption: Usage, limitations, and lessons learned. *Social Science Computer Review, 33*(5), 613–630.

Susha, I., Zuiderwijk, A., Janssen, M., & Grönlund, Å. (2015). Benchmarks for evaluating the progress of open data adoption: Usage, limitations, and lessons learned. *Social Science Computer Review, 33*(5), 613–630. https://doi.org/10.1177/0894439314560852

Swanson, E. B., & Ramiller, N. C. (1997). The organizing vision in information systems innovation. *Organization Science, 8*, 458–474.

Tammisto, Y., & Lindman, J. (2012). *Definition of open data services in software business*. Third international conference on software business. Cambridge, MA, USA

Teece, D. J. (2010). Business models, business strategy and innovation. *Long Range Planning, 43*(2–3), 172–194.

Tennison, J. (2012). *Open data business models*, retrievable from: http://www.jenitennison.com/2012/08/20/open-data-business-models.html

The World Bank. (2016). *GDP (current US$)*. Retrieved from https://data.worldbank.org/country/netherlands?view=chart

Torchiano, M., Vetro, A., & Iuliano, F. (2017). *Preserving the benefits of Open Government Data by measuring and improving their quality: an empirical study*. Paper presented at the Computer Software and Applications Conference (COMPSAC), 2017 IEEE 41st Annual.

Ubaldi, B. (2013a). *Open government data: Towards empirical analysis of open government data initiatives, OECD working papers on public governance, no 22* (p. 61). Paris, France: OECD Publishing https://doi.org/10.1787/5k46bj4f03s7-en

Ubaldi, B. (2013b). *Open government data: Towards empirical analysis of open government data initiatives*. Retrieved from Paris.

UK Cabinet Office. (2011). *Public Data Corporation to free up public data and drive innovation*. Retrieved from: https://www.gov.uk/government/news/public-data-corporation-to-free-up-public-data-and-drive-innovation

Umbrich, J., Neumaier, S., & Polleres, A. (2015). Towards assessing the quality evolution of open data portals.

United Arab Emirates – Federal Customs Authority. (2016). *Open data policy*. Retrieved from https://fca.gov.ae/en/pages/opendatapolicy.aspx?

Van de Does de Willebois, E., Halter, E., Harrison, R., Park, J., & Sharman, J. (2011). *The puppet masters: How the corrupt use legal structures to hide stolen assets and what to do about it*. Washington, DC: World Bank.

van de Walle, S. (2017). Trust in public administration and public services. In *Trust at Risk: Implications for EU Policies and Institutions* (pp. 118–128). Brussels, Belgium: European Union.

Van Loenen, B., Ubacht, J., Labots, W., & Zuiderwijk, A. (2017). *Log file analytics for gaining insight into actual use of open data*. Paper presented at the 17th European Conference on Digital Government, Lisbon, Portugal.

Van Veenstra, A. F., & van den Broek, T. A. (2013). *Opening moves. Drivers, enablers and barriers of open data in a semi-public organization*. Paper presented at the 12th Electronic Government Conference, Koblenz, Germany.

Venkatesh, V., & Bala, H. (2008). Technology acceptance model 3 and a research agenda on interventions. *Decision sciences, 39*(2), 273–315.

Venkatesh, V., & Davis, F. D. (2000). A theoretical extension of the technology acceptance model: Four longitudinal field studies. *Management Science, 45*(2), 186–204.

Venkatesh, V., & Zhang, X. (2010). Unified theory of acceptance and use of technology: US vs. China. *Journal of global information technology management, 13*(1), 5–27.

Venkatesh, V., Morris, M. G., Davis, G. B., & Davis, F. D. (2003). User acceptance of information technology: Toward a unified view. *MIS Quarterly, 27*(3), 425–478.

Vetrò, A., Canova, L., Torchiano, M., Minotas, C. O., Iemma, R., & Morando, F. (2016). Open data quality measurement framework: Definition and application to open government data. *Government Information Quarterly, 33*(2), 325–337.

Villazón-Terrazas, B., Vilches-Blázquez, L. M., Corcho, O., & Gómez-Pérez, A. (2011). Methodological guidelines for publishing government linked data. In D. Wood (Ed.), *Linking government data* (pp. 27–49). New York, NY: Springer.

Warner, J., & Chun, S. A. (2009). *Semantic and pragmatic annotation for government information discovery, sharing and collaboration*, Paper presented at the 10th annual international conference on digital government research (dg.o 2009), Puebla, Mexico — May 17 - 21, 2009.

Wass, C., Dini, P., Eiser, T., Heistracher, T., Lampoltshammer, T. J., Marcon, G., ... Winkels, R. (2013). OpenLaws.eu. In E. Schweighofer, F. Kummer, & W. Hötzendorfer (Eds.), *Abstraction and application: Proceedings of the 16th international legal informatics symposium* (Vol. 292, pp. 21–23). Vienna, Austria: Österreichische Computer Gesellschaft.

Weill, P., & Vitale, M. R. (2001). *Place to space: Migrating to e-business models*. Boston, MA: Harvard Business School Press.

Welle Donker, F., van Loenen, B., & Bregt, A. (2016). Open data and beyond. *International Journal of Geo-Information, 5*(4), 48. https://doi.org/10.3390/ijgi5040048

Welzel, C., Eckert, K.-P., Kirstein, F., & Jacumeit, V. (2017). *Mythos Blockchain: Herausforderung für den öffentlichen Sektor*. Berlin, Germany: Kompetenzzentrum Öffentliche IT - Fraunhofer-Institut für Offene Kommunikationssysteme FOKUS Retrieved from http://publica.fraunhofer.de/eprints/urn_nbn_de_0011-n-438569-19.pdf

Willcocks, L., & Graeser, V. (2001). *Delivering IT and E-business value*. Boston, MA: Butterworth–Heinemann.

Windrum, P., & Koch, P. (2008). *Innovation in public sector services. Entrepreneurship, creativity and management*. Celtenham, UK: Edward Elgar.

Wixom, B. H., & Todd, P. A. (2005). A theoretical integration of user satisfaction and technology acceptance. *Information Systems Research, 16*(1), 85–102.

World Bank. (2013a). *Open government data toolkit*. Available at: http://data.worldbank.org/ogd

World Bank. (2013b). *Open data readiness assessment tool*. Open Government Data Working Group. Retrieved from http://data.worldbank.org/sites/default/files/1/

World Bank Group. (2015). *Proposal for sustainable development goals*. World Bank. Retrieved from https://sustainabledevelopment.un.org/focussdgs.html

World Wide Web Consortium. (2014). *Data catalog vocabulary (DCAT)*. Retrieved from http://www.w3.org/TR/vocab-dcat/

World Wide Web Consortium. (2017). Data on the web best practices. *W3C Recommendation* 31 January 2017. Retrieved from http://www.w3.org/TR/dwbp/

World Wide Web Foundation. (2016). *Open data barometer*. Retrieved from http://opendatabarometer.org/

Yang, Z., & Kankanhalli, A. (2013). *Innovation in government services: The case of open data.* In Proceedings IFIPWG 8.6 international working conference on transfer and diffusion of IT, TDIT 2013 Banglore, India. pp. 644–651.

Yin, Y. (2017). *Video 3.3 – Privacy aspects of data sharing – Open data Governance: From policy to use.* Retrieved from https://www.youtube.com/watch?v=ZQMx7Uv6gPE&feature=youtu.be

Yu, H., & Robinson, D. G. (2012). The new ambiguity of 'open government'. *UCLA Law Review Discourse, 59*, 178–208. https://doi.org/10.2139/ssrn.2012489

Zeithaml, V. A. (2002). Service quality delivery through web sites: A critical review of extant knowledge. *Journal of the Academy of Marketing Science, 30*(4), 362–375.

Zeleti, F. A., Ojo, A., & Curry, E. (2014). Emerging business models for the open data industry: Characterization and analysis. *ACM* https://doi.org/10.1145/2612733.2612745

Zeleti, F. A., Ojo, A., & Curry, E. (2016). Exploring the economic value of open government data. *Government Information Quarterly, 33*(3), 535–551.

Zhao, L., & Ichise, R. (2014). Ontology integration for linked data. *Journal on Data Semantics, 3*(4), 237–254.

Zuiderwijk, A. (2015a). *Open data infrastructures: The design of an infrastructure to enhance the coordination of open data use.* 's-Hertogenbosch, The Netherlands: Uitgeverij BOXPress.

Zuiderwijk, A. (2015b). *Open data infrastructures: The design of an infrastructure to enhance the coordination of open data use.* (Doctoral Thesis), TU Delft, Delft.

Zuiderwijk, A. (Producer). (2016). *MOOC Open Government – Video 2.3 considerations when opening government data.* MOOC Open Government.

Zuiderwijk, A. (2017). Open data ProfEd – Video 2.3: Open data infrastructures. *Open data Governance: From policy to use.* Retrieved from https://online-learning.tudelft.nl/courses/open-data-governance-from-policy-to-use/

Zuiderwijk, A., Helbig, B., Gil-García, J. R., & Janssen, M. (2014). Special issue on innovation through open data – A review of the state-of-the-art and an emerging research agenda: Guest editors' introduction. *Journal of Theoretical and Applied Electronic Commerce, 9*(2.) Talca May 2014. https://doi.org/10.4067/S0718-18762014000200001

Zuiderwijk, A., & Janssen, M. (2013). *A coordination theory perspective to improve the use of open data in policy-making.* Paper presented at the 12th Conference on Electronic Government, Koblenz, Germany.

Zuiderwijk, A., & Janssen, M. (2014a). Open data policies, their implementation and impact: A framework for comparison. *Government Information Quarterly, 31*(1), 17–29 https://doi.org/10.1016/j.giq.2013.04.003

Zuiderwijk, A., & Janssen, M. (2014b). *The negative effects of open government data – Investigating the dark side of open data.* Paper presented at the Proceedings of the 15th Annual International Conference on Digital Government Research, Aguascalientes, Mexico.

Zuiderwijk, A., & Janssen, M. (2014c). *The negative effects of open government data – Investigating the dark side of open data.* Proceedings of the 15th annual international conference on digital government research, 2014, pp. 147–152. https://doi.org/10.1145/2612733.2612761

Zuiderwijk, A., & Janssen, M. (2015). Towards decision support for disclosing data: Closed or open data? *Information Polity, 20*(2, 3), 103–117.

Zuiderwijk, A., Janssen, M., Choenni, S., & Meijer, R. (2014). Design principles for improving the process of publishing open data. *Transforming Government: People, Process and Policy, 8*(2), 185–204. https://doi.org/10.1108/TG-07-2013-0024

Zuiderwijk, A., Janssen, M., Choenni, S., Meijer, R., & Alibaks, R. S. (2012). Socio-technical impediments of open data. *Electronic Journal of Electronic Government, 10*(2), 156–172.

Zuiderwijk, A., Janssen, M., & Dwivedi, Y. K. (2015). Acceptance and use predictors of open data technologies: Drawing upon the unified theory of acceptance and use of technology. *Government Information Quarterly, 32*(4), 429–440.

Zuiderwijk, A., Janssen, M., Meijer, R., Choenni, S., Charalabidis, Y., & Jeffery, K. (2012a). Issues and guiding principles for opening governmental judicial research data. In H. J. Scholl,

M. Janssen, M. Wimmer, C. Moe, & L. Flak, (Eds.), *Lecture notes in computer science (including subseries lecture notes in artificial intelligence and lecture notes in bioinformatics)*. Kristiansand, Norway. https://doi.org/10.1007/978-3-642-33489-4_8

Zuiderwijk, A., Janssen, M., Poulis, K., & van de Kaa, G. (2015). *Open data for competitive advantage: Insights from open data use by companies*. Paper presented at the Proceedings of the 16th Annual International Conference on Digital Government Research.

Zuiderwijk, A., Janssen, M., Van de Kaa, G., & Poulis, K. (2016). The wicked problem of commercial value creation in open data ecosystems: Policy guidelines for governments. *Information Polity, 21*(3), 223–236.

Zuiderwijk, A., Jeffery, K., & Janssen, M. (2012a). The potential of metadata for linked open data and its value for users and publishers. *JeDEM-eJournal of eDemocracy and Open Government, 4*(2), 222–244.

Zuiderwijk, A., Jeffery, K., & Janssen, M. (2012b). *The necessity of metadata for open linked data and its contribution to policy analyses*. In Conference on E-democracy and open government (CeDEM 2012) (pp. 281–294).

Zuiderwijk, A., Loukis, E., Alexopoulos, C., Janssen, M., & Jeffery, K. (2014). Elements for the development of an open data marketplace. In *Conference for e-democracy and open governement* (p. 309).

Appendix B: Abbreviations

API	Application Programming Interface
B2B	Business-to-Business
BC	Block-Chain
BCT	Blockchain Technology
BD	Big Data
BDA	Big Data Analytics
BOLD	Big, Open, Linked Data
CAPEX	Capital Expenditure
CERIF	Common European Research Information Format
CGD	Citizen-Generated Data
CKAN	Comprehensive Knowledge Archive Network
CRIS	Current Research Information Systems
CS	Civil Servant
CSV	Comma-Separated Values
DC	Dublin Core
DCAT	Data Catalog Vocabulary
DDI	Data Documentation Initiative
DDoS attack	Distributed Denial of Service attack
DQ	Data Quality
DWBR-WG	Data on the Web Best Practices Working Group
EC	European Commission
EDP	European Data Portal
e-GMS	e-Government Metadata Standard
EGRL	E-Government Reference Library
ESA	European Space Agency
ETL	Extract, Transform, Load task
FOAF	Friend-of-a-Friend Ontology
FOIA	Freedom of Information Act
GPS	Global Positioning System
ID	Identity Document

Y. Charalabidis et al., *The World of Open Data*, Public Administration and Information Technology 28, https://doi.org/10.1007/978-3-319-90850-2

IDS	Industrial Data Space
INSPIRE	Infrastructure for Spatial Information in the European Community
IoT	Internet of Things
IRI	Internationalized Resource Identifier
IS	Information Systems
ISO	International Organization for Standardization
I(C)T	Information (and Communication) Technologies
JSON	JavaScript Object Notation
KPI	Key Performance Indicator
LD	Linked Data
LOD	Linked Open Data
LODC	Linked Open Data Cloud
LOSD	Linked Open Statistical Data
OBD	Open Business Data
OD	Open Data
ODI	Open Data Initiative
ODP	Open Data Portal
OGD	Open Government Data
OGD	Open Government Data
OGP	Open Government Partnership
OPEX	Operational Expenditure
OWL	Web Ontology Language
OWMS	Overheid.nl Web Metadata Standard
P2P	Peer-to-Peer
PA	Public Administration
PEM	Privacy-Enhancement Mechanisms
PEOU	Perceived Ease-of-Use
PPDP	Privacy-Preserving Data Publishing
PSI	Public Sector Information
PU	Perceived Usefulness
RDF	Resource Description Framework
RDF	Resource Description Framework
REST	Representational State Transfer
RFID	Radio-Frequency Identification
RIF	Rule Interchange Format
ROI	Return on Investment
SDI	Scientific Data Infrastructure
SDMX	Statistical Data and Metadata eXchange
SKOS	Simple Knowledge Organization System
SPARQL	RDF Query Language
SQuaRE	Software Product Quality Requirements and Evaluation
SWOT	Strengths, Weaknesses, Opportunities, Threats
TAM	Technology Acceptance Model
TFP	Total Factor Productivity
TPB	Theory of Planned Behaviour

TRA	Theory of Reasoned Action
TURTLE	Terse RDF Triple Language
URI	Unique Resource Identifier
URL	Unique Resource Locator
UTAUT	Unified Theory of Acceptance and Use of Technology
W3C	World Wide Web Consortium
XLS	Microsoft Excel File Format

Appendix C: Index

Y. Charalabidis et al., *The World of Open Data*, Public Administration and Information Technology 28, https://doi.org/10.1007/978-3-319-90850-2

Appendix D: Author Biographies

Yannis Charalabidis, Associate Professor, University of the Aegean
Blog: www.charalabidis.gr, Twitter: @yannisc

Yannis Charalabidis is Associate Professor in the Department of Information and Communication Systems Engineering at the University of Aegean. In parallel, he serves as Director of the Innovation and Entrepreneurship Unit of the University, designing and managing youth entrepreneurship activities, and Head of Information Systems Laboratory, coordinating policy making, research and pilot application projects for governments and enterprises worldwide. He has more than 20 years of experience in designing, implementing, managing and applying complex information systems as project manager, in Greece and Europe. He has been employed for 8 years as an executive director in SingularLogic Group, leading software development and company expansion in Greece, Eastern Europe, India and the USA. He has published more than 150 papers in international journals and conferences, while actively participating in international standardization committees and scientific bodies. In 2014 he was nominated as the eighth most productive writer in the world, among 9500 scholars in the Electronic Government domain, according to the Washington University survey. He is Best Paper Award winner in the International IFIP e-Government Conference (2008, 2012), winner of the first prize in OMG/Business Process Modelling contest (2009) and second prize winner in the European eGovernment Awards (2009).

Y. Charalabidis et al., *The World of Open Data*, Public Administration and Information Technology 28, https://doi.org/10.1007/978-3-319-90850-2

Anneke Zuiderwijk, Post-doctoral researcher, Delft University of Technology
Twitter: @AnnekeZuiderwyk, @OpendataX

Dr. Anneke Zuiderwijk is a post-doctoral researcher in the Information and Communication Technology research group of the Faculty of Technology, Policy, and Management at Delft University of Technology, the Netherlands. She obtained her PhD with distinction (only awarded to the top 5% of TU Delft PhD candidates) for her research on the design of open data infrastructures and she was one of four award nominees for the most talented female PhD candidate at TU Delft (out of 101 candidates). Anneke was ranked as one of the most influential open data researchers worldwide (Hossain, Dwivedi & Rana, 2015[1]) and has also been a recipient of the international Digital Governance Junior Scholar Award. Anneke's research now focuses on what drives different categories of researchers from different disciplines to open up and use research data or not.

Anneke is a track chair at the EGOV-CeDEM-ePart Conference, chairing the track 'Open Data, Linked Data and Semantic Web'. She is also a Programme Committee member at several conferences/workshops, including CeDEM, Dg.o, EGOV, ICEGOV and ICEDEG. She co-chairs the PhD colloquium of the EGOV-CeDEM-ePart conference and is leading two work packages in the VRE4EIC project (A Europe-wide interoperable Virtual Research Environment to Empower multidisciplinary research communities and accelerate Innovation and Collaboration).

Charalampos Alexopoulos, Post-doctoral Researcher, University of the Aegean

Charalampos Alexopoulos is a post-doctoral researcher in the Department of Information and Communications Systems Engineering at the University of the Aegean, publishing at several conferences and in journals on open data, decision support, smart cities and e-government. He is also a researcher in the Information Systems Laboratory of the same department, working on European and national funded research and pilot application projects (ENGAGE, SHARE-PSI 2.0, EU-COMMUNITY, PADGETS, NOMAD, NET-EUCEN, PLUG-IN) for

[1] Hossain, M. A., Dwivedi, Y. K., & Rana, N. P. (2016). State-of-the-art in open data research: Insights from existing literature and a research agenda. Journal of Organizational Computing and Electronic Commerce, 26(1-2), 14-40.

governments and enterprises. Charalampos is a computer science graduate from the University of Peloponnese with an MSc in Management Information Systems and PhD on open data infrastructures from the University of the Aegean. He serves as programme and organization committee member of the annual Samos Summit on ICT-enabled Governance and he is a minitrack co-chair at the HICSS conference series. His teaching activities include the pre-graduate course of eGovernment, the post-graduate course of Management Information Systems being also the course manager of the 'Open & Collaborative Governance' and 'Big Data Analysis on Earth Sciences' summer schools. In 2015, Charalampos was ranked as one of the most prolific researchers in open data research by Hossain, Dwivedi and Rana (2015) and he has won first and second place in the WeGov Awards of 2010 and 2011. Charalampos is currently the project manager of the Gov 3.0 project 'Scientific Foundations, Training and Entrepreneurship Activities in the Domain of ICT-Enabled Governance'.

Marijn Janssen, Professor, Delft University of Technology

Marijn Janssen is Full Professor in ICT and Governance and Head of the Information and Communication Technology section of the Technology, Policy and Management Faculty of Delft University of Technology. He is Co-Editor-in-Chief of *Government Information Quarterly*, Associate Editor of the *International Journal of Electronic Business Research* (IJEBR), *Electronic Journal of eGovernment* (EJEG), *International Journal of E-Government Research* (IJEGR), *Decision Support Systems* (DSS) and *Information Systems Frontiers* (ISF). He serves as conference chair of IFIP EGOV series, programme chair of IFIP I3E2016, track chair at AMCIS and minitrack chair at the HICSS conference series. His research is focused on the design and service orchestration of public-private service networks. Public-private networks can be characterized by interacting public and private parties having different objectives and requirements, various degrees of technology-readiness, a plurifom systems landscape, path dependencies and the need to be compliant with the regulatory environment. Service orchestration is aimed at integrating disparate activities performed by separated organizations taking into account aspects ranging from the institutional and organizational level to the technical level. By current technology developments like cloud computing, software as a service, semantic services, linked open data and policy developments like open data this landscape is fundamentally changing. The traditional relationship between governments and the public is challenged resulting in a more open government. He was ranked as one of the leading e-government researchers in a survey in 2009 and 2014 and has published over 320 refereed publications.

Thomas Lampoltshammer, Assistant Professor, Danube University Krems, Austria

Dr. Thomas Lampoltshammer is Assistant Professor for ICT & Governance and Deputy Head of the Centre for E-Governance in the Department of E-Governance and Administration at the Danube University Krems/Austria. Prior to his current position, he worked as a researcher and lecturer in Applied Informatics in the School of Information Technology and Systems Management at the Salzburg University of Applied Sciences Salzburg/Austria. His research interests include, but are not limited to, Geoinformatics, Semantics, Open Data, Data Visualization and Software Engineering. His project experience as PI includes EU-funded research projects and national grants in the domain of data management, organizational theory and ICT in public administration. He is co-founder and co-chair of the International Data Science Conference (iDSC) series, as well as member of the ICA Commission on Cognitive Issues in Geographic Information Visualization and the Institute of Electrical and Electronics Engineers (IEEE). Besides his academic activities within the International Conference for E-Democracy and Open Government (CeDEM), he acts as reviewer for several international conferences as well as for numerous SCI-indexed journals. His current research focus is in the domain of open data, the effects of ICT application in a connected society and the effects on a data-driven society.

Enrico Ferro, PhD, Head of Department, ISMB
Blog: www.enricoferro.com – Twitter: @egferro

Enrico Ferro is Head of the Innovation Development Area at Mario Boella Institute (ISMB), a multidisciplinary team of researchers working on the strategic, socio-economic and policy implications of information and communication technologies. Over the last 15 years Dr. Ferro has worked in many projects financed by the European Commission with roles ranging from scientific supervisor to senior expert. Dr. Ferro also covers an Adjunct Professor position at the International Labour Bureau of the United Nations where he regularly lectures on innovation management in the public sector. His research work has produced over 40 academic publications, 1 handbook of research and over 50 research reports.

The manufacturer's authorised representative in the EU is Springer Nature Customer Service Centre GmbH, Europaplatz 3, 69115 Heidelberg, Germany. If you have any concerns regarding our products, please contact ProductSafety@springernature.com

Printed and bound by CPI Group (UK) Ltd, Croydon, CR0 4YY
30/09/2026
02225662-0003